Student Solutions Manual
to accompany

THE ANALYSIS AND DESIGN OF LINEAR CIRCUITS & LAPLACE EARLY

FOURTH EDITION

Roland E. Thomas
Professor Emeritus
United States Air Force Academy

Albert J. Rosa
Professor and Chairman of Engineering
University of Denver

Prepared By:
Ronald R. DeLyser
Professor of Engineering
University of Denver

WILEY
JOHN WILEY & SONS, INC.

Cover Photo Credit: Rob Atkins/The Image Bank/Getty Images

To order books or for customer service call 1-800-CALL-WILEY (225-5945).

ISBN 0-471-46968-8

Printed in the United States of America

10 9 8 7 6 5 4 3 2 1

Printed and bound by Hamilton Printing Company

Preface

We have created this Student Solutions Manual (SSM) in response to the many requests by students who wanted to have available worked-out examples to serve as additional solutions of typical Circuits problems beyond those found in the text. This supplementary mini-text solves problems from <u>The Analysis and Design of Linear Circuits</u>, by Thomas and Rosa, fourth edition – both the Traditional and the Laplace Early versions. The SSM uses a variety of readily available and current software tools to solve the various problems. For example, MATLAB, Mathcad and Microsoft Excel are used for mathematical manipulations and graphing, while Orcad Capture and Electronics Workbench are used for simulations of circuits. In most cases the selection of the proper tool for the problem solution has been made. In some very rare cases, the tool selected did not work well so another, more appropriate tool was selected. This Student Solutions Manual discusses those cases.

In preparing this manual we have used Microsoft Word to present all solutions except for those using Mathcad. Screen shots of schematics and some graphs are embedded in the Word documents. The majority of MATLAB problems have been solved using m-files. The major reason is that mistakes can be easily corrected in the m-file editor. Results shown at the command line have been edited, removing the excessive number of spaces and line feeds. In presenting the outputs from Orcad Capture's Probe the Output File has been edited to show only the pertinent data.

We have concentrated as much as possible on the use of the computer tool, rather than hand analysis. For completeness, we have also included the analytical solutions necessary, the logic to design approaches, and any necessary preliminary work to the use of computer tools. In several cases we solved the same problem using different software tools and have commented on the results.

The Table of Content lists the chapter, by version when appropriate, from which the problem was taken, the software tool or tools used, and the page on which the solution can be found. Every problem begins on a new page.

The text authors welcome suggestions to improve this manual.

The basic Student Solutions Manual was prepared by Dr. Ronald R. DeLyser for the third edition and updated for the current edition by Drs. Albert J. Rosa and Roland E. Thomas.

Table of Contents

1-8 Figure P1-8 shows a plot of the net positive charge flowing in a wire versus time. Sketch the corresponding current during the same time period.

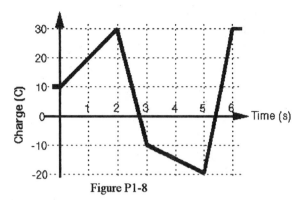

Figure P1-8

Solution: We can make use of the programming feature of Mathcad. To see how this is done, select Mathcad Help under the Help menu (or use the shortcut key F1) under the Contents tab, double click on "Programming". Double click on "Defining a program". You will then see step by step directions on how to define the charge function. (Of course you need your general math skills to define the piecewise linear function from the graph.)

$$q(t) := \begin{cases} (10 + 10 \cdot t) & \text{if } 0 \le t < 2 \\ [\, 30 - 40 \cdot (t - 2)\,] & \text{if } 2 \le t < 3 \\ [\,-10 - 5 \cdot (t - 3)\,] & \text{if } 3 \le t < 5 \\ [\,-20 + 50 \cdot (t - 5)\,] & \text{if } 5 \le t < 6 \\ 30 & \text{if } 6 \le t \end{cases}$$

Then use the derivative function to get the current function.

$$i(t) := \frac{d}{dt} q(t)$$

Finally, define the range of the independent variable, t, and graph both functions on an XY plot.

$$t := 0, 0.01 .. 6$$

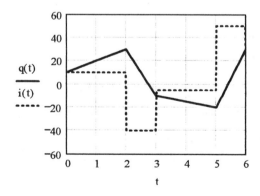

1-11 For $t \leq 5$ s the current through a device is $i(t) = 10$ A. For $t > 5$ s it is $i(t) = 20 - 2t$ A. Sketch $i(t)$ versus time and find the total charge entering the device between $t = 0$ s and $t = 10$ s.

Solution: The following is an m-file for use with MATLAB. Notice that comments are an essential part of any program and are included in this solution.

```
% Problem 1-11, Solution
% First we define the time period.
t=0:0.1:10;
% Next, we define the current function for the two time periods.
t1=0:0.1:5;
i(1:length(t1))=10;
t2=5.1:0.1:10;
i((length(t1)+1):(length(t1)+length(t2)))=20-2*t2;
% Next, the current function is plotted.
plot(t,i(1:length(t)))
xlabel('sec')
ylabel('i(t) (Amps)')
% Define some symbols for the anticipated symbolic integration.
syms t Q
% Do the integration of the current function.
Q=int(10,t,0,5)+int(20-2*t,t,5,10)
```

After running the m-file, a figure is produced in another window. It contains the desired graph required by the problem statement and is shown below.

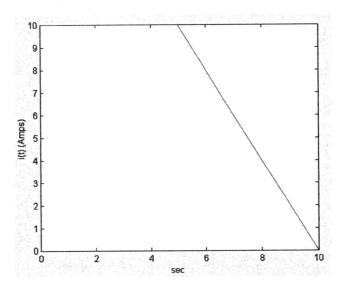

Figure 1-11s. The plot of current versus time.

The definite integral is evaluated resulting in a total charge of $Q = 75$. This is a fairly simple integral to evaluate and MATLAB is used primarily for illustration of use of the Symbolic Toolbox. Let's do the integral analytically and check the results.

$$Q = \int_0^5 10 \, dt + \int_5^{10} 20 - 2t \, dt = 10t \Big|_0^5 + 20t - t^2 \Big|_5^{10}$$
$$= 50 + 20(10 - 5) - (100 - 25) = 75$$

1-26 POWER RATIO IN dB

In complete darkness the voltage across and current through a two-terminal light detector are +5.6 V and +8 nA. In full sunlight the voltage and current are +0.9 V and +4 mA. Express the light/dark power ratio of the device in decibels (dB).

We will make use of MATLAB for this exercise. The m-file with comments follows:

```
% Problem 1-26
% In complete darkness the voltage across and current through
% a two-terminal light detector are +5.6 V and +8 nA.  In full
% sunlight the voltage and current are +0.9 V and +4 mA.
% Express the light/dark power ratio of the device in decibels (dB).
%
% First we define the voltages and currents for dark and light conditions:
Vdark=5.6;
Idark=8*10^(-9);
Vlight=0.9;
Ilight=4*10^(-3);
% Now we calculate the powers associated with dark and light conditions:
Pdark = Vdark*Idark
Plight = Vlight*Ilight
% The power ratio in dB is finally:
PdB = 10*log10(Plight/Pdark)
%
```

MATLAB gives the following results after the m-file is executed:

```
Pdark = 4.4800e-008
Plight = 0.0036
PdB = 49.0502
```

2.6 The *i-v* characteristic of a nonlinear resistor is $v = 75i + 0.2i^2$.

(a) Calculate v and p for i = ±0.5, ±1, ±2, ±5, and ±10 A.

(b) If the operating range of the device is limited to |i| < 0.5 A, what is the maximum error in v when the device is approximated by a 100 Ω linear resistance?

We will use Microsoft Excel for this problem. The columns for *i, v, p, R, iR* and the error are shown below.

I Amperes	v Volts	p Watts	R Ohms	iR Volts	Error Percent
-10	-730.00	7300.00			
-5	-370.00	1850.00			
-2	-149.20	298.40			
-1	-74.80	74.80			
-0.5	-37.45	18.73	100	-50	25.1
0	0.00	0.00	100	0	0
0.5	37.55	18.78	100	50	24.9
1	75.20	75.20			
2	150.80	301.60			
5	380.00	1900.00			
10	770.00	7700.00			

4

2-9 Figure P2-9 shows the circuit symbol for a two-terminal device called a *diode*.

Figure P2-9

For a *p-n* junction diode, theoretical analysis yields the following *i-v* relationship:

$$i = 2 \times 10^{-16} \left(e^{40v} - 1 \right)$$

(d) Use this equation to find *i* and *p* for *v* = 0, ±0.1, ±0.2, ±0.4, and ±0.8 V. Use these data to plot the *i-v* characteristic of the element.

We will use MATCAD to solve the equation for the specified values of *v* and to plot *i(v)* and *p(v)*. The following setup is used:

$$v := -0.8, -0.7 .. 0.8$$

$$i(v) := 2 \cdot 10^{-16} \left(e^{40v} - 1 \right)$$

The resulting graph is shown below in Figure P2-9s.

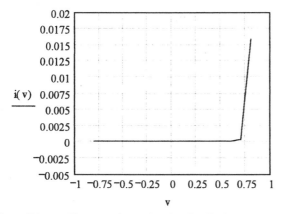

Figure P2-9s. The *i-v* and *p-v* plots for the diode.

(b) Is the diode linear or nonlinear, bilateral or nonbilateral, and active or passive?

From the graph of the *i-v* characteristics, it is nonlinear, nonbilateral and passive.

(d) Use the diode model to predict *i* and *p* for *v* = 5 V. Do you think the model applies to voltages in this range? Explain.

The same setup as above was used for this calculation. The results were:

$v := 5$

$i(v) = 1.445 \times 10^{71}$

$p := v \cdot i(v)$

$p = 7.226 \times 10^{71}$

These values are unreasonable for current and power. The device would be destroyed.

(d) Repeat (c) for $v = -5$ V.

The results from MATCAD for this calculation are

$v := -5$

$i(v) = 0$

$p := v \cdot i(v)$

$p = 1 \times 10^{-15}$

These values are reasonable since they are well within the operating range of these devices.

2-22 First use KVL to find the voltage across each resistor in Figure P2-22. Then use these voltages and KCL to find the current through every element, including the voltage sources.

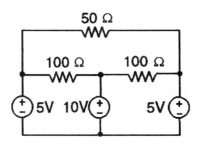

Figure P2-22

KVL 0utermost loop:	$-5 + V_6 + 5 = 0$	$V_6 = 0$	$i6 = 0$
KVL lower right loop:	$-10 + V_5 + 5 = 0$	$V_5 = 5$	$i5 = 5V/100\Omega = 50\ mA$
KVL lower left loop:	$-5 + V_4 + 10 = 0$	$V_4 = -5$	$i4 = -5V/100\Omega = -50\ mA$
KCL left node:	$i_1 - i_4 - i_6 = 0$	$i_1 = i_4$	$i_1 = -50\ mA$
KCL right node:	$i_3 + i_5 + i_6 = 0$	$i_3 = -i_5$	$i_3 = -50\ mA$
KCL center node:	$i_2 + i_4 - i_5 = 0$	$i_2 = i_5 - i_4$	$i_2 = 100\ mA$

We now check our work with a circuit simulator, Electronics Workbench. The DC simulation was done with the proper components and voltage and current indicators. The circuit and indicated voltages and currents are shown in Figure P2-22s. Notice that the simulator gives very small values for V_6 and I_6 instead of 0.

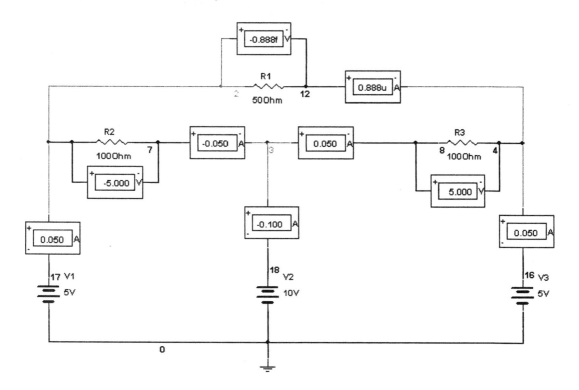

Figure P2-22s. Electronics Workbench simultor results.

2-47 Figure P2-47 shows an ammeter circuit consisting of a D'Arsonval meter, a two-position selector switch, and two shunt resistors. A current of 0.5 mA produces full-scale deflection of the D'Arsonval meter, whose internal resistance is $R_M = 50\ \Omega$. Select the shunt resistance R_1 and R_2 so that $i_x = 10$ mA produces full-scale deflection when the switch is in position A and $i_x = 50$ mA produces full-scale deflection when the switch is in position B.

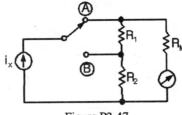

Figure P2-47

Solution: For all conditions we have the full-scale current and meter resistance given by

$$i_{FS} = 0.5\ \text{mA} \qquad R_M = 50\ \Omega$$

When the switch is in position A, the desired current is

$$i_{xA} = 10\ \text{mA}.$$

Using current division we have

$$(1) \qquad i_{FS} = \frac{(R_1 + R_2)i_{xB}}{R_1 + R_2 + R_M} \qquad \Rightarrow \qquad R_1 + R_2 = \frac{R_M i_{FS}}{i_{xA} - i_{FS}}$$

When the switch is in position B, the desired current is

$$i_{xB} = 50\ \text{mA}.$$

Using current division we have

$$(2) \qquad i_{FS} = \frac{R_2 i_{xA}}{R_1 + R_2 + R_M} \qquad \Rightarrow \qquad R_2 = \frac{(R_1 + R_2 + R_M)i_{FS}}{i_{xA}}$$

Once we calculate $R_1 + R_2$ from (1), we can calculate R_2 from (2). R_2 can then be substituted into (1) to give R_1. We do this with the following m-file in MATLAB.

```
% The two circuits (switch in position A, and switch in position B) are
% analyzed and expressions for R2 and R1 + R2 are found.  We solve for R1
% and R2 here.
%
% First we define the full scale current and resistance of the D'Arsonval
meter.
%
IFS=0.0005;
RM=50;
% The desired values if ix for positions A and B are
ixA=0.01;
ixB=0.05;
% The value of R1 + R2 is given by
R1plusR2=(RM*IFS)/(ixA-IFS);
```

8

```
% The value of R2 is given by
R2=(RM+R1plusR2)*IFS/ixB
% Finally, R1 is calculated from
R1=R1plusR2-R2
%
```

The answers are:
```
R2 = 0.5263
R1 = 2.1053
```

2-61   (A) DEVICE MODELING

The circuit in Figure P2-61 consists of a 50-Ω linear resistor in parallel with a nonlinear *varistor* whose *i-v* characteristic is $i_V = 2.6 \times 10^{-5}v^3$.
(a) Plot the *i-v* characteristic of the parallel combination.
(b) State whether the parallel combination is linear or nonlinear, active or passive, and bilateral or nonbilateral.
(c) Identify a range of voltages over which the parallel combination can be modeled within ±10% by a linear resistor.
(d) Identify a range of voltages over which the parallel combination can be safely operated if both devices are rated at 50 W. Which device limits this range?
(e) The parallel combination is connected in series with a 50-Ω resistor and a 5-V voltage source. In this circuit, how would you model the parallel combination and why?

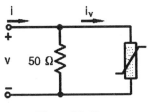

Figure P2-61

Solution:

(a) Using Microsoft *Excel*, the following table is generated where the V column is the input, the I column uses the equation:

=2.6*(10^(-5))*A2^3+A2/50

for the sum of the currents through the *varistor* and the 50 Ω resistor, and the I50 column is the current through the 50 Ω resistor.

Table P2-61a

V	I	I50
-100	-28	-2
-90	-20.754	-1.8
-80	-14.912	-1.6
-70	-10.318	-1.4
-60	-6.816	-1.2
-50	-4.25	-1
-40	-2.464	-0.8
-30	-1.302	-0.6
-20	-0.608	-0.4
-10	-0.226	-0.2
0	0	0
10	0.226	0.2
20	0.608	0.4
30	1.302	0.6
40	2.464	0.8
50	4.25	1
60	6.816	1.2
70	10.318	1.4
80	14.912	1.6
90	20.754	1.8
100	28	2

10

Columns 1 and 2 of the table are used to produce the following graph:

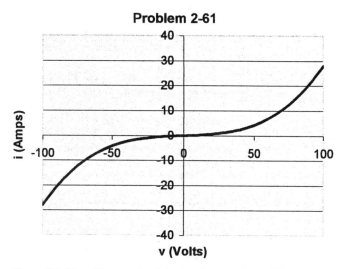

Problem 2-61

Figure P2-61sa. The graph of the *i-v* characteristic of Figure P-61.

(b) The parallel combination is nonlinear (*i-v* curve is not a straight line), passive (*i-v* curve lies in the first and third quadrants so that *vi* > 0), and bilateral (*i-v* curve has odd symmetry about the origin).

(c) The answer to this question is somewhat complicated. From Table P2-61, it is clear that in the range of *-10 < v < 10*, the total current is approximately the same as that through the 50 Ω resistor. If we model the parallel combination with a 50 Ω resistor and compare the true current with the 50 W resistor current, we get the following:

Table P2-61b

V	I	Ilin	%Error	R
-15	-0.38775	-0.3	-29.25	50
-14	-0.35134	-0.28	-25.48	
-13	-0.31712	-0.26	-21.97	
-12	-0.28493	-0.24	-18.72	
-11	-0.25461	-0.22	-15.73	
-10	-0.226	-0.2	-13	
-9	-0.19895	-0.18	-10.53	
-8	-0.17331	-0.16	-8.32	
-7	-0.14892	-0.14	-6.37	
-6	-0.12562	-0.12	-4.68	
-5	-0.10325	-0.1	-3.25	
-4	-0.08166	-0.08	-2.08	
-3	-0.0607	-0.06	-1.17	
-2	-0.04021	-0.04	-0.52	
-1	-0.02003	-0.02	-0.13	
0	0	0	0	
1	0.020026	0.02	-0.13	
2	0.040208	0.04	0	
3	0.060702	0.06	-1.17	
4	0.081664	0.08	-2.08	
5	0.10325	0.1	-3.25	
6	0.125616	0.12	-4.68	

11

7	0.148918	0.14	-6.37
8	0.173312	0.16	-8.32
9	0.198954	0.18	-10.53
10	0.226	0.2	-13
11	0.254606	0.22	-15.73
12	0.284928	0.24	-18.72
13	0.317122	0.26	-21.97
14	0.351344	0.28	-25.48
15	0.38775	0.3	-29.25

The percent error (column 4) is less than 10% for an input voltage range of approximately $-9 < v < 9$. However, if we change the size of the resistor (column 5) incrementally and keep the range of voltages around 0 V below $\pm$ 10 %, we find (see table P2-61c) that we can model the parallel combination by a 45 Ω resistor over a range of greater than $-13 < v < 13$.

Table P2-61c

V	I	Ilin	%error	R
-15	-0.38775	-0.33333	-16.325	45
-14	-0.35134	-0.31111	-12.932	
-13	-0.31712	-0.28889	-9.773	
-12	-0.28493	-0.26667	-6.848	
-11	-0.25461	-0.24444	-4.157	
-10	-0.226	-0.22222	-1.7	
-9	-0.19895	-0.2	0.523	
-8	-0.17331	-0.17778	2.512	
-7	-0.14892	-0.15556	4.267	
-6	-0.12562	-0.13333	5.788	
-5	-0.10325	-0.11111	7.075	
-4	-0.08166	-0.08889	8.128	
-3	-0.0607	-0.06667	8.947	
-2	-0.04021	-0.04444	9.532	
-1	-0.02003	-0.02222	9.883	
0	0	0	0	
1	0.020026	0.022222	9.883	
2	0.040208	0.044444	0	
3	0.060702	0.066667	8.947	
4	0.081664	0.088889	8.128	
5	0.10325	0.111111	7.075	
6	0.125616	0.133333	5.788	
7	0.148918	0.155556	4.267	
8	0.173312	0.177778	2.512	
9	0.198954	0.2	0.523	
10	0.226	0.222222	-1.7	
11	0.254606	0.244444	-4.157	
12	0.284928	0.266667	-6.848	
13	0.317122	0.288889	-9.773	
14	0.351344	0.311111	-12.932	
15	0.38775	0.333333	-16.325	

(d) The portion of the *Excel* sheet with this answer is shown below:

P	Vvar	Vres
50	37.2391	50

12

The P column is the input (cell A27). The Vvar column is the voltage across the *varistor* given the power. The equation in the cell is

=(A27/0.000026)^0.25

The Vres column is the voltage across the resistor for the given power. The cell equation is

=SQRT(50*A27)

The results show that the varistor limits the safe range to -37.239 V to +37.239 V.

(e) If modeled as a 45 Ω resistor, then by voltage division (one cell in the *Excel* sheet) the voltage across the modeled resistor would be 2.368 V. If modeled as a nonlinear element we need to solve two equations - one for the total current (I) through the parallel combination and the other for the sum of the voltages across the parallel combination (V) and the 50 Ω resistor. The two equations are

(1) $\qquad I = \dfrac{V}{50} + 2.6 \times 10^{-5} V^3$

(2) $\qquad 5 = 50I + V$

Solving (2) for V and substituting this into (1) gives the following cubic equation in V:

(3) $\qquad 2.6 \times 10^{-5} V^3 + \dfrac{V}{25} - \dfrac{5}{50} = 0$

We can put this equation into a cell (third column in Table P2-61d is cell C36, =2.6*10^(-5)*B36^3+B36/25-0.1) and put guess values in another referenced cell (second column in Table P2-61d which is cell B36).

Table P2-61d

Veq	Vguess	Vnl	%Error45	%Error50
2.368421	2.48999	9.98E-07	4.882314	-0.402

By changing the guess we can get the equation as close to zero as we want (depending on how much time we want to spend). But, there is an easier way - *Excel* will do this for us. Under the Tools menu item, choose Solver. A dialog box will pop up. Figure P2-61sb shows the box after it has been filled in. The target cell is the cell with the cubic equation (cell C36) referring to the guess value (cell B36). We want this value to be equal to 0 so we select "Value of:" and put 0 in the place holder. We then want to change the guess value so we select that cell (cell B36). We click solve and another dialog will appear asking us if we want to accept the solution. Click OK and we are done.

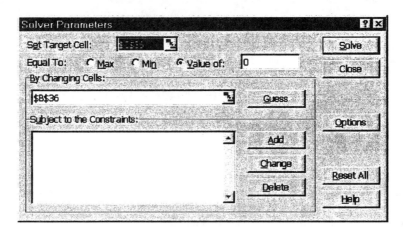

Figure P2-61sb. Dialog box for the solution of Equation (3).

Table P2-61d also shows results for the error between the linear model (45 Ω resistor) and the exact model which is 4.88 %. If we had used a 50 Ω linear model (see discussion in (c) above) the voltage across it would be 2.5 V giving an error of 0.402 %, also shown in Table P2-61d. So, even though the 45 Ω resistor provides a linear model over a larger range (given the 10 % error constraint) than the 50 Ω resistor, the 50 Ω resistor provides a more accurate model for this circuit constraint.

3-13 (a) Formulate mesh-current equations for the circuit in Figure P3-13.

(b) Solve for v_x and i_x using $R_1 = 10$ kΩ, $R_2 = 10$ kΩ, $R_3 = 2$ kΩ, $R_4 = 1$ kΩ, $i_S = 2.5$ mA, $v_{S1} = 12$ V, and $v_{S2} = 0.5$ V.

(c) Find the power Supplied by v_{S1}.

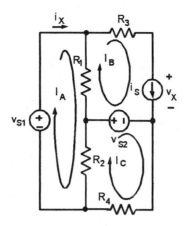

Figure P3-13

(a) Mesh Equations:

Mesh A: $-V_{S1} + R_1(I_A-I_B) + R_2(I_A-I_C) = 0$

Mesh B: $I_B=I_S$

Mesh C: $V_{S2} + R_4(I_C) + R_2(I_C-I_A) = 0$

(b) The problem is simulated in Orcad Capture Lite Edition with the schematic shown in Figure P3-13s. The simulation gives voltages and currents everywhere, but you have the option to delete any you don't need. This leaves the desired values of 3.350 mA for i_x and 7 V - 3 V = 4 V for v_x. For your information, all of the Mesh currents and Node Voltages have been left.

(c) The power from source v_{S1} is P = 12 x 3.350 = -40.2 mW.

(d)

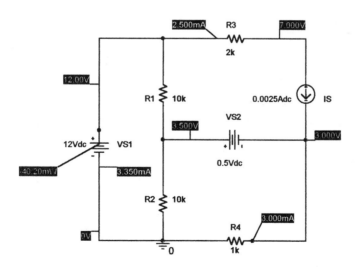

Figure P3-13s. Orcad Capture schematic and output values.

3-14 The circuit in Figure P3-14 seems to require two supermeshes since both current sources appear in two meshes. However, a circuit diagram can sometimes be rearranged to eliminate the need for supermesh equations. (a) Show that supermeshes in Figure P3-14 can be avoided by connecting resistor R_6 between node A and node D via a different route.
(b) Formulate mesh-current equations for the modified circuit as redrawn in (a).
(c) Solve for v_x using $R_1 = R_2 = R_3 = R_4 = 2$ kΩ, $R_5 = R_6 = 1$ kΩ, $i_{S1} = 40$ mA, and $i_{S2} = 20$ mA.

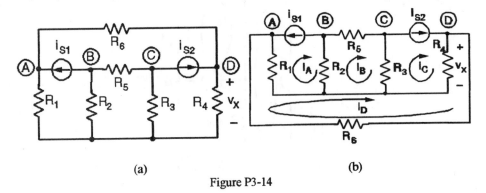

(a)   (b)

Figure P3-14

(a) Figure P3-14(b) shows the redrawn circuit that does not require supermeshes to solve.
(b) Mesh A: $I_A = -I_{S1}$.
     Mesh B: $R_2(I_B - I_A) + R_5(I_B) + R_3(I_B - I_C) = 0$
     Mesh C: $I_C = I_{S2}$
     Mesh D: $R_6(I_D) + R_1(I_D - I_A) + R_4(I_D - I_C) = 0$

(c) We will do this part of this problem in Electronics Workbench. The schematic is shown in Figure P3-14s. The indicated value for v_x is 55.933 V.

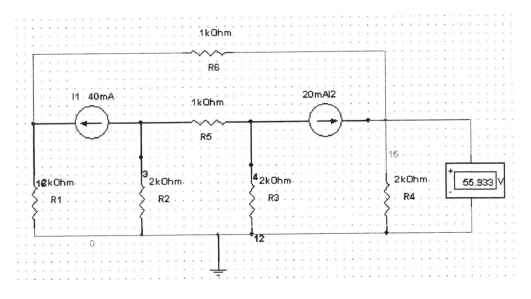

Figure P3-14s

3-29 Use the Superposition principle in the circuit of Figure P3-29 to find i_0.

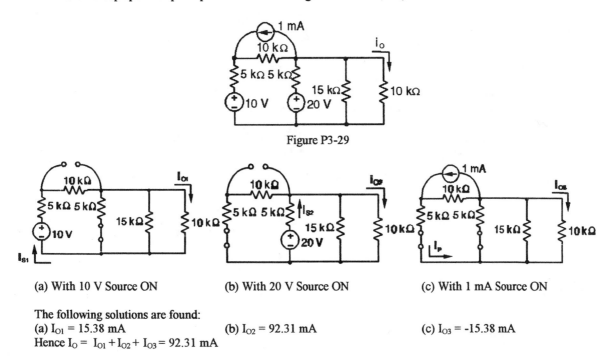

Figure P3-29

(a) With 10 V Source ON

(b) With 20 V Source ON

(c) With 1 mA Source ON

The following solutions are found:
(a) I_{O1} = 15.38 mA
(b) I_{O2} = 92.31 mA
(c) I_{O3} = -15.38 mA
Hence $I_O = I_{O1} + I_{O2} + I_{O3}$ = 92.31 mA

We can validate this solution by use Orcad Capture. The Orcad Capture schematic along with the value for i_0 = 923.1 µA (and many other values) is shown in Figure P3-29s.

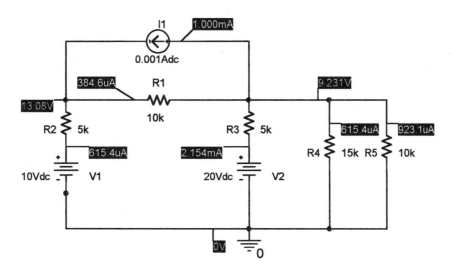

Figure P3-29s

17

3-44 Some measurements of the *i-v* characteristic of a two-terminal source are as follows:

v (V)	-10	-5	0	5	10	12	13	14
i (ma)	5	4	3	2	1	0	-1	-2

(a) Plot the source *i-v* characteristic.
(b) Develop a Thévenin equivalent circuit valid on the range $|v| < 10$ V.
(c) Use the equivalent circuit to predict the source v_{OC} and i_{SC}.
(d) Compare your results in (c) with the measurements given and explain any differences.

We use the following m-file to plot the source *i-v* characteristic, calculate the Thévenin equivalent circuit parameters, and predict the source v_{OC} and i_{SC} using the equivalent circuit.

```
% First we create the i and v vectors from the data given in the problem
statement.
%
i=[.005,.004,.003,.002,.001,0,-.001,-.002];
v=[-10 -5 0 5 10 12 13 14];
% A plot for these values is easily generated.
%
plot(v,i);
xlabel('v (Volts)');
ylabel('i (Amps)');
%
% For the given range, the i-v characteristic is a straight line with slope
%
m=(i(5)-i(1))/(v(5)- v(1));
%
% with i axis intercept of
b=i(3);
%
% so that the equation for this line is i = m*v + b.  The i-v characteristic
% of a Norton equivalent is i = -1/RT * v + iN.  Hence for the given range
% the parameters of the Norton equivalent are
iN=b;
RT=-1/m;
% and the Thevenin voltage is
%
vT=iN*RT;
%
% The source open circuit voltage and short circuit current are simply
vOC=vT
iSC=iN
%
```

The output from this m-file is:

vOC = 15
iSC = 0.0030

and the plot is shown in Figure P3-44.

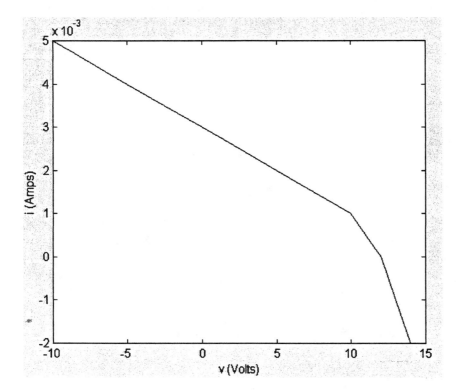

Figure P3-44

3-61 Figure P3-61 shows a two-port interface circuit connecting source and load circuits. In this problem $v_S = 10$ V, $R_S = 50\ \Omega$, and the load is a $50\ \Omega$ resistor. To avoid damaging the source, its output current must be less than 100 mA. Design a resistive interface circuit so that the voltage delivered to the load is 4 V and the source current is less than 100 mA.

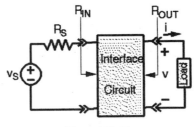

Figure P3-61

We use Microsoft Excel to compute the values for this problem. The following Excel output shows the calculation of load current (80 mA) for the required 4 V load voltage.

RS	Vs	RL	VL	io=VL/RL	iS=io
50	10	50	4	0.08	0.08

This value of current is less than the specified maximum of 100 mA. Clearly, we do not want to increase the current from the source very much more than this value. Also, since the required voltage is less than half of the source voltage, it seems reasonable to create our interface circuit with a simple resistor in series with R_S and the load. The voltage across this resistor, $v_R = v_S - v_L - i_S R_S$, is calculated along with the required resistance, $R = v_R/i_S$, below.

VR	R
2	25

The final schematic is simulated with Electronic WorkBench verifying the validity of the design. The results are shown in Figure P3-61s.

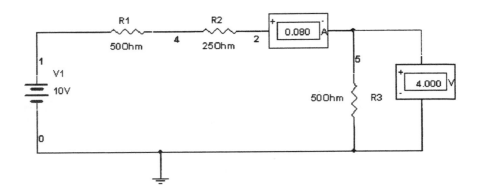

Figure 3-61s

3-64 **(D)** In Figure P3-61 the load is a 500 Ω resistor and R_S = 75 Ω. Design an interface ₍
the input resistance of the two-port is 75 Ω ±10% and the output resistance seen by the loa₍

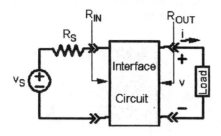

Figure P3-61

Solution: Since the input resistance is less than the load resistance, we need the interface circuit to be in parallel with the load. The output resistance is greater than the source resistance, dictating a series interface circuit. We choose R_1 and R_2 as the interface circuit, resulting in Figure P3-61s as the complete circuit which was drawn in Electronics Workbench.

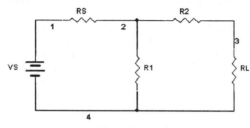

Figure P3-61s

The given values are:

$$R_S := 75 \cdot \Omega \qquad\qquad R_L := 500 \cdot \Omega$$

The design constraints are:

$$R_{in} := 75 \cdot \Omega \qquad\qquad R_{out} := 500 \cdot \Omega$$

The design equations will be given in a solve block with initial guesses

$$R_1 := 75 \cdot \Omega \qquad\qquad R_2 := 500 \cdot \Omega$$

Given

$$R_{out} = R_2 + \frac{R_1 \cdot R_S}{R_1 + R_S} \qquad\qquad R_{in} = \frac{R_1 \cdot (R_2 + R_L)}{R_1 + R_2 + R_L}$$

$$\begin{pmatrix} R_1 \\ R_2 \end{pmatrix} := Find(R_1, R_2) \qquad\qquad \begin{pmatrix} R_1 \\ R_2 \end{pmatrix} = \begin{pmatrix} 81.349 \\ 460.977 \end{pmatrix} \Omega$$

If we use standard value 5 % resistors of values 82 Ω and 470 Ω, we should meet the 10 % accuracy constraint.

21

3-68 (D) Figure P3-61 shows a two-port interface circuit connecting a source and load. In this problem the source with $v_S = 5$ V and $R_S = 5$ Ω is to be used in production testing of two-terminal semiconductor devices. The devices are to be connected as the load in Figure P3-61 and have highly nonlinear and variable i-v characteristics. The normal operating range for acceptable devices is $\{i > 10$ mA or $v > 0.7$ V$\}$ and $\{p < 10$ mW$\}$. Design an interface circuit so that the operating point always lies within the specified normal range regardless of the test article's i-v characteristic.

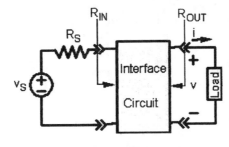

Figure P3-61

Solution: We will use Excel to chart the allowable region of safe operation. A column of input voltage over a range of 0 V to 2 V with a column of calculated values of I for a power of 10 mW gives the power curve. Another column of .01 Amp up to .7 Volt then dropping to 0 Amp, gives the minimum values of the voltage and current. The resulting plot is shown in Figure P3-61s. The allowable range of load lines is in the region of the plots between the power curve (upper trace of Figure P3-61s) and the minimum values of v and i (lower rectangular region of Figure P3-61s..

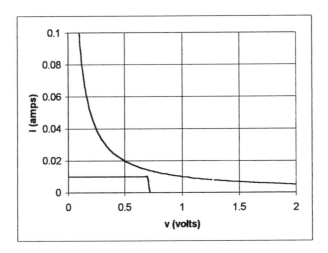

Figure P3-61sa. Allowable range of load lines for the interface circuit.

For our proposed load line, we can choose a point midway between the minimum current (.01 Amp) and the value of current at the minimum voltage (.7 V) for the maximum power (.014 Amp taken from the Excel spreadsheet). We choose the point $(V_1, I_1) = (.7, .012)$ for one point on the load line. For one other point, we choose $(V_2, I_2) = (0, 0.025)$. This gives a line with a slope of $-(.025-.012)/0.7$ and an i intercept of .025. The equation of this line is then

$$i = -\frac{0.013}{0.7}v + 0.025$$

Using the Excel spreadsheet, we plot this line along with the previous data. The resulting graph is shown in Figure P3-61sb. Of course, this is only one possible load line for this design. The design parameters for this load line are $i_{sc} = 25$ mA, $v_{oc} = 1.346$ V, and $R_T = 53.85\ \Omega$.

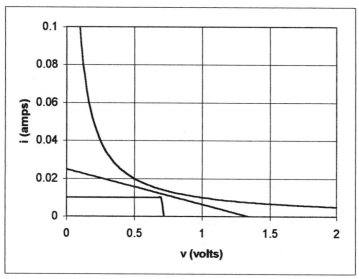

Figure P3-61sb. One possible load line for the design.

3-75   (E)   TTL TO ECL CONVERTER

It is claimed that the resistive circuit in Figure P3-75 converts transistor-transistor logic (TTL) input signals into output signals compatible with emitter coupled logic (ECL). Specifically, the claim is that for any output current in the range -0.025 mA $\leq i \leq$ 0.025 mA, the circuit converts any input in the TTL low range ($0 \leq v_S \leq 0.4$ V) to an output in the ECL low range (-1.7 V $\leq v \leq$ -1.5 V), and converts any input in the TTL high range ($3.0 \leq v_S \leq 3.8$ V) to an output in the ECL high range (- 0.9 V $\leq v \leq$ -0.6 V). The purpose of this problem is to verify this claim.

(a) The output voltage can be written in the form

$$v = k_1\, v_S + k_2\, i + k_3$$

Write a KCL equation at the output with the current i as an unknown and solve for the constants k_1, k_2 and k_3.

(b) Use the relationship found in (a) to verify that output voltage falls in one of the allowed ECL ranges for every allowed combination of TTL inputs and load currents.

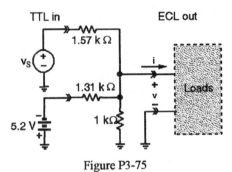

Figure P3-75

(a)   The KCL equation at the node common to the three resistors is

$$\frac{v_S - v}{1.57\,k\Omega} + \frac{-5.2 - v}{1.32\,k\Omega} - \frac{v}{1\,k\Omega} = i$$

We can work at the command line in MATLAB to solve this equation for v in terms of vs and i. Using the Solve function we write

```
solve('(vs-v)/1570+(-5.2-v)/1320-v/1000=i','v')
```

The result is

```
ans =

.26600032242463324197968724810576*vs-1.64517169111720135418345961 63147-
417.62050620667418990810897952604*i
```

We truncate to three significant figures to get

$$v(v_S, i) = 0.266\,v_S - 417\,i - 1.64$$

so that $k_1 = 0.266$, $k_2 = -417$ and $k_3 = -1.64$.

(b)   We create the following m-file in MATLAB (the m-file makes it easier than the command line because of the edit features of the m-file editor).

```
vs=0
i=0.025*10^-3
```

```
v=.266*vs-417*i-1.64
vs=0.4
i=0.025*10^-3
v=.266*vs-417*i-1.64
vs=0
i=-0.025*10^-3
v=.266*vs-417*i-1.64
vs=0.4
i=-0.025*10^-3
v=.266*vs-417*i-1.64
vs=3.0
i=0.025*10^-3
v=.266*vs-417*i-1.64
vs=3.8
i=0.025*10^-3
v=.266*vs-417*i-1.64
vs=3.0
i=-0.025*10^-3
v=.266*vs-417*i-1.64
vs=3.8
i=-0.025*10^-3
v=.266*vs-417*i-1.64
```

The results after running the m-file and which verify the design are:

vs = 0
i = 2.5000e-005
v = -1.6504

vs = 0.4000
i = 2.5000e-005
v = -1.5440

vs = 0
i = -2.5000e-005
v = -1.6296

vs = 0.4000
i = -2.5000e-005
v = -1.5232

vs = 3
i = 2.5000e-005
v = -0.8524

vs = 3.8000
i = 2.5000e-005
v = -0.6396

vs = 3
i = -2.5000e-005
v = -0.8316

vs = 3.8000
i = -2.5000e-005
v = -0.6188

4-2 For the circuit in Figure P4-2,
  (a) Find the voltage gain v_O/v_1.
  (b) Find the current gain i_O/i_S.
  (c) For $i_S = 2$ mA, find the power supplied by the input source i_S and the power delivered to the 2-kΩ load resistor.

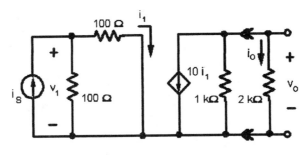

Figure P4-2

We will use Electronics Workbench to solve this problem. Figure P4-2s shows the schematic entered into Electronics Workbench and the results for some select currents and voltages.

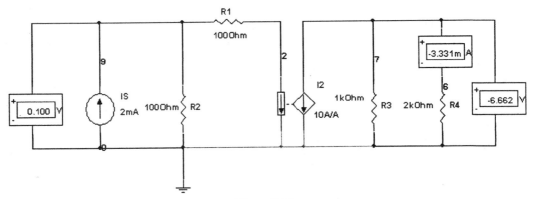

Figure P4-2s

  (a) The voltage gain, v_O/v_1 is calculated to be -6.662 V/0.1 V = -66.62.
  (b) The current gain, i_O/i_1 is calculated to be -3.331 mA/2 mA = -1.67.
  (c) The power supplied by the current source is PS = .1 V × 2 mA = 0.2 mW. The power delivered to the load is PL = 6.662 V × 3.331 mA = 22.19 mW.

4-3 The circuit in Figure P4-3 is a dependent-source model of a two-stage amplifier. For $v_S = 1$ mV, find the output voltage v_3 and the current gain i_3/i_1.

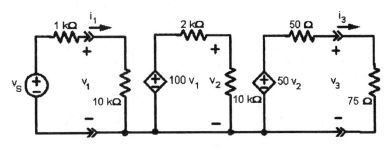

Figure P4-3

We will use Orcad Capture to work this problem. The schematic and simulation results are shown in Figure P4-3s.

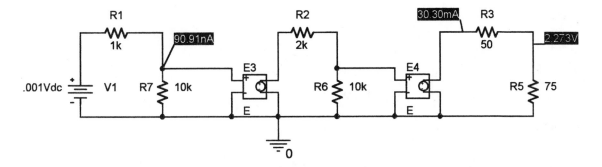

Figure 4-3s

The output voltage, v_3 is equal to 2.273 V, and the current gain $i_3/i_1 = 30.3$ mA/ 90.91 nA = 333300.

4-8 The circuit in Figure P4-8 is a model of a feedback amplifier using two identical transistors. Formulate either node-voltage or mesh-current equations for this circuit. Use these equations to solve for the input-output relationship $v_O = Kv_S$ using $r_x = 1\ k\Omega$, $R_E = 200\ \Omega$, $R_C = 10\ k\Omega$, $R_L = 5\ k\Omega$, $R_F = 5\ k\Omega$, and $\beta = 100$.

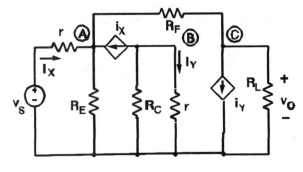

Figure P4-8

Solution: The node equations are:

Node A: $\quad \dfrac{v_A}{200} + \dfrac{v_A - v_O}{5000} - 100\dfrac{v_S - v_A}{1000} - \dfrac{v_S - v_A}{1000} = 0$

Node B: $\quad \dfrac{v_B}{10000} + \dfrac{v_B}{1000} + 100\dfrac{v_S - v_A}{1000} = 0$

Node C: $\quad 100\dfrac{v_B}{1000} + \dfrac{v_O}{5000} + \dfrac{v_O - v_A}{5000} = 0$

We can use these equations and the Symbolic Toolbox in MATLAB. The m-file with $v_S = 1$ follows:

```
% Problem 4-8
%
% First we have to define the symbols.
%
syms vA vB vO
%
% Next, the equations, f = 0.
%
f1=vA/200+(vA-vO)/5000-100*(1-vA)/1000-(1-vA)/1000;
f2=vB/10000+vB/1000+100*(1-vA)/1000;
f3=10*vB/10+vO/5000+(vO-vA)/5000;
%
% Finally, use the solve function to get the node voltages.  Because we set
% v1 = 1, the value of vO is the voltage gain of the circuit.
%
[vA,vB,vO]=solve(f1,f2,f3);
K=v0
%
```

The result is:

K = 43335185/1670557

or K = 25.94

4-20 (D) The input source in Figure P4-20 is a series connection of a dc source V_{BB} and a signal source v_S. The circuit parameters are $R_B = 500\ k\Omega$, $R_C = 5\ k\Omega$, $\beta = 100$, $V_\gamma = 0.7$ V, and $V_{CC} = 15$ V.
(a) With $v_S = 0$, select the value of V_{BB} so that the transistor is in the active mode with $v_{CE} = V_{CC}/2$.
(b) Using the value of V_{BB} found in (a), find the range of values of the signal voltage v_S for which the transistor remains in the active mode.
(c) Plot the transfer characteristic v_{CE} versus v_S as the signal voltage sweeps across the range from -10 V to + 10 V.

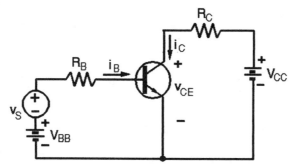

Solution: In order to solve this in MATHCAD we first we need to make the definitions from the problem statement.

$$R_B := 500 \cdot k\Omega \qquad R_C := 5 \cdot k\Omega \qquad \beta := 100 \qquad V_\gamma := 0.7 \cdot V \qquad V_{CC} := 15 \cdot V$$

(a) With $v_S = 0$ and the given design parameters we have

$$v_{CE} := \frac{V_{CC}}{2} \qquad i_C := \frac{V_{CC} - v_{CE}}{R_C} \qquad i_B := \frac{i_C}{\beta} \qquad V_{BB} := i_B \cdot R_B + V_\gamma$$

$$v_{CE} = 7.5\ V \qquad i_C = 1.5\ mA \qquad i_B = 15\ \mu A \qquad V_{BB} = 8.2\ V$$

(b) For the transistor to be in the active mode

$$i_B = \frac{v_S + V_{BB} - V_\gamma}{R_B} > 0 \qquad \text{so that} \qquad v_S > V_\gamma - V_{BB}$$

$$V_\gamma - V_{BB} = -7.5\ V$$

$$i_C = \beta \cdot \left(\frac{v_S + V_{BB} - V_\gamma}{R_B}\right) < \frac{V_{CC}}{R_C} \quad \text{so that} \qquad v_S < V_\gamma - V_{BB} + \frac{V_{CC} \cdot R_B}{\beta \cdot R_C}$$

$$V_\gamma - V_{BB} + \frac{V_{CC} \cdot R_B}{\beta \cdot R_C} = 7.5\ V$$

So, for the transistor to be in the active mode, $-7.5\ V < v_S < 7.5\ V$.

(c) For the range of $-10\ V < v_S < 10$ V, we have three regions of operation. The transistor is cutoff for $-10\ V < v_S < -7.5$ V, it is in the linear region for $-7.5\ V < v_S < 7.5$ V, and it is saturated for $7.5\ V < v_S < 10$ V. We can describe this transfer function using the programming feature of Mathcad. To get to the programming toolbar, under the View menu item, select Toolbars and Programming. After you make the hit the assign key (:), select add line in the Programming toolbar. A vertical line with two place holders will appear. The rest is up to you.

29

$$v_{CE}(v_S) := \begin{cases} V_{CC} & \text{if } v_S < -7.5 \cdot V \\[2ex] V_{CC} - \beta \cdot \dfrac{v_S + V_{BB} - V_\gamma}{R_B} \cdot R_C & \text{if } -7.5 \cdot V \le v_S \le 7.5 \cdot V \\[2ex] 0 & \text{if } v_S > 7.5 \end{cases}$$

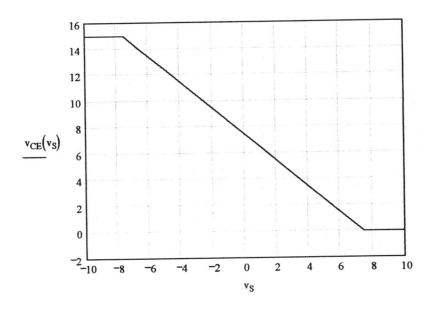

30

4-33 For the OP AMP circuit in Figure P4-33, find the output v_O in terms of the input v_S.

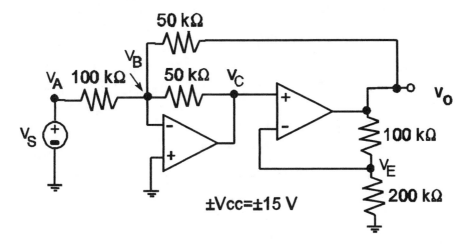

Figure P4-33

We can simulate this in Electronics Workbench as long as we keep the simulation in the linear region of the OP AMPs. The circuit is shown below.

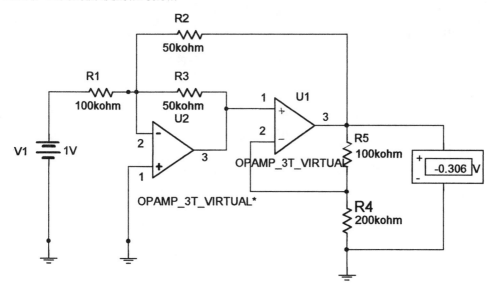

Figure P4-33s

Since the input is 1 V and the output is -0.306 V, the transfer function of the circuit is $v_O = -.306\ v_S$. If power supplies of $\pm$ 15 volts are used for the OP AMPs, the output indicates that they are operating in the linear region.

A hand calculation shows that $v_O = -0.3\ v_S$.

4-37 (D) Design circuits using resistors and OP AMPs to implement each of the following input-output relationships:

(a) $v_O = 3v_1 - 2v_2$

(b) $v_O = 2v_1 + v_2$

Solution:

(a) For this transfer function, the simplest solution is to use a differential amplifier (shown in Figure 4.37sa).

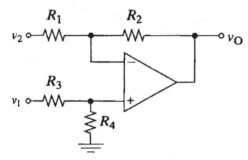

Figure 4.37sa

The transfer function for this circuit is

$$v_O = \frac{R_1 + R_2}{R_1} \cdot \frac{R_4}{R_3 + R_4} \cdot v_1 - \frac{R_2}{R_1} \cdot v_2 \quad \text{with} \quad \frac{R_1 + R_2}{R_1} \cdot \frac{R_4}{R_3 + R_4} = 3 \text{ and } \frac{R_2}{R_1} = 2$$

Since there are four unknown resistors, but only two gain equations, we need to pick two resistors. For the gain on the inverting side, we choose

$$R_1 := 10 \cdot k\Omega \qquad \text{so that} \qquad R_2 := 2 \cdot R_1$$

$$R_2 = 20 \, k\Omega$$

With these values, we have

$$\frac{R_1 + R_2}{R_1} = 3$$

so that

$$\frac{R_4}{R_3 + R_4} = 1$$

The only value for R_3 that will satisfy this equation will be $R_3 = 0 \, \Omega$. The value of R_4 can be selected to be anything we want. Constraints for the selection of R4 would be the input impedance or balancing the resistance seen by the inverting and non-inverting terminals of the OP AMP. We'll use this last constraint for the selection of R_4.

$$R_4 := \frac{R_1 \cdot R_2}{R_1 + R_2} \qquad\qquad R_4 = 6.667 \times 10^3 \, \Omega$$

This completes the design

32

(b) For the second transfer function, we use a weighted summer followed by an inverter (shown in Figure 4-37sb).

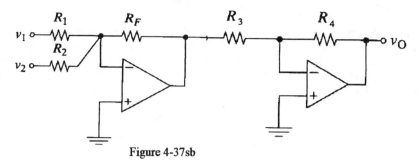

Figure 4-37sb

The expression for this transfer function is

$$v_O = \frac{-R_4}{R_3} \cdot \left(\frac{-R_F}{R_1} \cdot v_1 - \frac{R_F}{R_2} \cdot v_2 \right) = \frac{R_4 \cdot R_F}{R_3 \cdot R_1} \cdot v_1 + \frac{R_4 \cdot R_F}{R_3 \cdot R_2} \cdot v_2 \text{ with } \quad \frac{R_4 \cdot R_F}{R_3 \cdot R_1} = 2 \quad \text{and} \quad \frac{R_4 \cdot R_F}{R_3 \cdot R_2} = 1$$

We'll choose $R_3 := 10 \cdot k\Omega$ $\quad R_4 := 10 \cdot k\Omega$ $\quad R_F := 10 \cdot k\Omega$ so that we are left with calculating

$$R_1 := \frac{R_F}{2} \qquad\qquad R_1 = 5\,k\Omega$$

$$R_2 := R_F \qquad\qquad R_2 = 10\,k\Omega$$

This completes the design.

33

4-43 (D) A requirement exists for an amplifier with a gain of -12,000 and an input resistance of at least 300 kΩ. Design an OP AMP circuit that meets the requirements using general-purpose OP AMPs with voltage gains of $A = 2\times10^5$, input resistances of $R_I = 4\times10^8$ Ω, and output resistances $R_O = 20$ Ω.

Solution: We need to make the closed-loop gains of each stage less than 1 % of the open-loop gain of 200,000 or less than 2000. We can do this with two stages with gains of 120 and -100. Since the requirement is for an input resistance of at least 300 kΩ, if we have the first stage be the non-inverting stage with gain of 120, the input resistance will be much greater than 300 kΩ. The second stage is then an inverting OP AMP stage with gain 100. The circuit is shown in Figure P4-43.

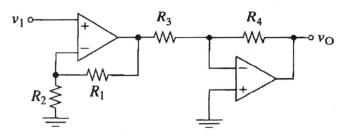

Figure P4-43

For the values of the resistors, we first choose $R_4 = 10$ kΩ so that $R_3 = 100$ kΩ. For the non-inverting stage, we choose $R_2 = 10$ kΩ and use the command line of MATLAB to solve for R_1. For the symbolic processor, we need the two equations f1 = 0 and f2 = 0. The equations are for the gain of the non-inverting amplifier and for the value of R_2. The result is:

```
syms R1 R2
f1=120-(R1+R2)/R2;
f2=10000-R2;
[R1,R2]=solve(f1,f2)

R1 =

1190000

R2 =

10000
```

We will use the standard resistor size of 1.2 MΩ in place of the 1.19 MΩ called for in the solution.

4-58 (A, D, E) PHOTORESISTOR INSTRUMENTATION

Both circuits in Figure P4-58 contain a photoresistor R_X whose resistance varies inversely with the intensity of the incident light. In complete darkness its resistance is 10 kΩ. In bright sunlight its resistance is 2 kΩ. At any given light level the circuit is linear, so its input-output relationship is of the form $v_O = Kv_1$.

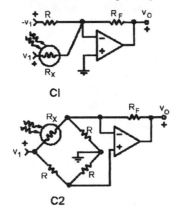

CI

C2

Figure P4-58

(a) **(A)** For circuits C1 and C2 determine the constant K in terms of circuit resistances.
(b) **(D)** For circuit C1 with $v_1 = +15$ V, select the values of R and R_F so that $v_O = -10$ V in bright sunlight and +10 V in complete darkness.
(c) **(D)** Repeat part (b) for Circuit C2.
(d) **(E)** Compare these two designs on the basis of the number of devices required and the total power dissipated in each design.

Solution:
(a) Circuit C1 is a half-bridge amplifier which operates as an inverting summer. The transfer function is

$$v_O = \frac{-R_F}{R} \cdot (-v_1) - \frac{R_F}{R_X} \cdot v_1 = R_F \cdot \left(\frac{1}{R} - \frac{1}{R_X} \right) \cdot v_1 = K \cdot v_1 \qquad K = R_F \cdot \left(\frac{1}{R} - \frac{1}{R_X} \right)$$

Circuit C2 is a full bridge amplifier. The node equations at the OP AMP inverting and non-inverting inputs respectively are:

$$\left(\frac{1}{R_X} + \frac{1}{R} + \frac{1}{R_F} \right) \cdot v_N - \frac{v_1}{R_X} - \frac{v_O}{R_F} = 0 \qquad \frac{2 \cdot v_P}{R} - \frac{v_1}{R} = 0 \quad \text{or} \quad v_P = \frac{v_1}{2}$$

For and ideal OP AMP, $v_P = v_N = v_1/2$. Substituting this into the inverting input equations gives

$$\left(\frac{1}{R_X} + \frac{1}{R} + \frac{1}{R_F} \right) \cdot \frac{v_1}{2} - \frac{v_1}{R_X} - \frac{v_O}{R_F} = 0$$

Solving for v_0 gives

$$v_O = R_F \left(\frac{1}{R_F} + \frac{1}{R} - \frac{1}{R_X} \right) \frac{v_1}{2} \qquad K = \frac{1}{2} \left(1 + \frac{R_F}{R} - \frac{R_F}{R_X} \right)$$

35

(b) In bright sunlight and complete darkness we define the values R_{Xs} and R_{Xd} as the values of R_X respectively.

$$R_{Xs} := 2 \cdot k\Omega \qquad\qquad R_{Xd} := 10 \cdot k\Omega$$

For $v_1 := 15 \cdot V$ we solve the following solve block for circuit C1 with guesses $R := 1 \cdot k\Omega \quad R_F := 1 k\Omega$

Given

$$-10 \cdot V = R_F \cdot \left(\frac{1}{R} - \frac{1}{R_{Xs}} \right) \cdot v_1 \qquad\qquad 10 \cdot V = R_F \cdot \left(\frac{1}{R} - \frac{1}{R_{Xd}} \right) \cdot v_1$$

$$\begin{pmatrix} R_{C1} \\ R_{FC1} \end{pmatrix} := \text{Find}(R, R_F) \qquad\qquad \begin{pmatrix} R_{C1} \\ R_{FC1} \end{pmatrix} = \begin{pmatrix} 3.333 \\ 3.333 \end{pmatrix} k\Omega$$

(c) We have the same input information for circuit C2. We can even use the same values of R and R_F guesses for circuit C2. The equations we need to solve this time are

Given

$$-10 \cdot V = \frac{1}{2} \left(1 + \frac{R_F}{R} - \frac{R_F}{R_{Xs}} \right) \cdot v_1 \qquad\qquad 10 \cdot V = \frac{1}{2} \left(1 + \frac{R_F}{R} - \frac{R_F}{R_{Xd}} \right) \cdot v_1$$

$$\begin{pmatrix} R_{C2} \\ R_{FC2} \end{pmatrix} := \text{Find}(R, R_F) \qquad\qquad \begin{pmatrix} R_{C2} \\ R_{FC2} \end{pmatrix} = \begin{pmatrix} 6.667 \\ 6.667 \end{pmatrix} k\Omega$$

(d) Let's calculate the power dissipated by the external resistors. For C1, the most power dissipated is under full sunlight conditions (R_X is minimum) and is given by

$$v_O := -10 \cdot V$$

$$P_{C1} := \frac{(-v_1)^2}{R_{C1}} + \frac{v_1{}^2}{R_{Xs}} + \frac{(v_O)^2}{R_{FC1}} \qquad\qquad P_{C1} = 0.21 \, W$$

For C2, we note that the potential at the inverting and non-inverting terminals is $v_1/2$. The power dissipated under full sunlight conditions is then

$$P_{C2} := \left(\frac{v_1}{2} \right)^2 \cdot \frac{R_{C2} + R_{Xs}}{R_{C2} \cdot R_{Xs}} + \left(\frac{v_1}{2} \right)^2 \cdot \frac{2}{R_{C2}} + \frac{\left(v_O - \frac{v_1}{2} \right)^2}{R_{FC2}} \qquad P_{C2} = 0.099 \, W$$

Circuit C2 requires 2 additional resistors but dissipates less power than circuit C1.

4-60 (D) COMPUTER-AIDED CIRCUIT DESIGN

Use computer-aided circuit analysis to find the value of R_F in Figure P4-60 that causes the input resistance seen by i_S to be 50 Ω. Find the current gain i_O/i_S for this value of R_F. Use $\beta = 100$, $r_\pi = 1.1$ kΩ, $R_C = 10$ kΩ, R_E, and $R_L = 100$ Ω.

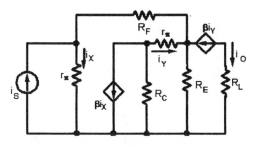

Figure P4-60

The schematic from Electronics Workbench shown in Figure P3-60s was used for this problem.

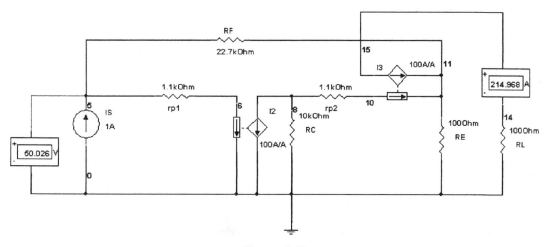

Figure p4-60s

The voltage indicator was monitored while the value of R_F was changed. A 50 Ω input impedance seen by the 1 A source would display 50 V on the indicator. The resulting current gain for the correct value of R_F is then displayed on the output current indicator and is shown to be 215.

5-6 Determine the amplitude and time constant of each of the following exponential waveforms. Graph each of the waveforms.

(a) $v_1(t) = [10e^{-2t}] u(t)$
(b) $v_2(t) = [10e^{-t/2}] u(t)$
(c) $v_3(t) = [-10e^{-20t}] u(t)$
(d) $v_4(t) = [-10e^{-t/20}] u(t)$

Solution: The form of the exponential waveform is

$$v(t) = \left[V_A e^{-t/T_C} \right] u(t)$$

where V_A is the amplitude and T_C is the time constant. By inspection, the amplitudes and time constants for the given waveforms are:

(a) 10, 0.5
(b) 10, 2
(c) -10, 0.05
(d) -10, 20

We use Excel to plot the waveforms. A rule of thumb is to plot for a time period of 0 to 5 T_C. The following chart is designed to have 50 points for each time scale (columns labeled t_A, t_B, t_C, and t_D). This is done by incrementing the time by a factor $5T_C/50$. The columns labeled A, B, C and D are the functions given in (a), (b), (c), and (d). The chart follows:

t_A	A	t_B	B	t_C	C	t_D	D
0	10	0	10	0	-10	0	-10
0.05	9.048374	0.2	9.048374	0.005	-9.04837	2	-9.04837
0.1	8.187308	0.4	8.187308	0.01	-8.18731	4	-8.18731
0.15	7.408182	0.6	7.408182	0.015	-7.40818	6	-7.40818
0.2	6.7032	0.8	6.7032	0.02	-6.7032	8	-6.7032
0.25	6.065307	1	6.065307	0.025	-6.06531	10	-6.06531
0.3	5.488116	1.2	5.488116	0.03	-5.48812	12	-5.48812
0.35	4.965853	1.4	4.965853	0.035	-4.96585	14	-4.96585
0.4	4.49329	1.6	4.49329	0.04	-4.49329	16	-4.49329
0.45	4.065697	1.8	4.065697	0.045	-4.0657	18	-4.0657
0.5	3.678794	2	3.678794	0.05	-3.67879	20	-3.67879
0.55	3.328711	2.2	3.328711	0.055	-3.32871	22	-3.32871
0.6	3.011942	2.4	3.011942	0.06	-3.01194	24	-3.01194
0.65	2.725318	2.6	2.725318	0.065	-2.72532	26	-2.72532
0.7	2.46597	2.8	2.46597	0.07	-2.46597	28	-2.46597
0.75	2.231302	3	2.231302	0.075	-2.2313	30	-2.2313
0.8	2.018965	3.2	2.018965	0.08	-2.01897	32	-2.01897
0.85	1.826835	3.4	1.826835	0.085	-1.82684	34	-1.82684
0.9	1.652989	3.6	1.652989	0.09	-1.65299	36	-1.65299
0.95	1.495686	3.8	1.495686	0.095	-1.49569	38	-1.49569
1	1.353353	4	1.353353	0.1	-1.35335	40	-1.35335
1.05	1.224564	4.2	1.224564	0.105	-1.22456	42	-1.22456
1.1	1.108032	4.4	1.108032	0.11	-1.10803	44	-1.10803
1.15	1.002588	4.6	1.002588	0.115	-1.00259	46	-1.00259
1.2	0.90718	4.8	0.90718	0.12	-0.90718	48	-0.90718
1.25	0.82085	5	0.82085	0.125	-0.82085	50	-0.82085
1.3	0.742736	5.2	0.742736	0.13	-0.74274	52	-0.74274

1.35	0.672055	5.4	0.672055	0.135	-0.67206	54	-0.67206
1.4	0.608101	5.6	0.608101	0.14	-0.6081	56	-0.6081
1.45	0.550232	5.8	0.550232	0.145	-0.55023	58	-0.55023
1.5	0.497871	6	0.497871	0.15	-0.49787	60	-0.49787
1.55	0.450492	6.2	0.450492	0.155	-0.45049	62	-0.45049
1.6	0.407622	6.4	0.407622	0.16	-0.40762	64	-0.40762
1.65	0.368832	6.6	0.368832	0.165	-0.36883	66	-0.36883
1.7	0.333733	6.8	0.333733	0.17	-0.33373	68	-0.33373
1.75	0.301974	7	0.301974	0.175	-0.30197	70	-0.30197
1.8	0.273237	7.2	0.273237	0.18	-0.27324	72	-0.27324
1.85	0.247235	7.4	0.247235	0.185	-0.24724	74	-0.24724
1.9	0.223708	7.6	0.223708	0.19	-0.22371	76	-0.22371
1.95	0.202419	7.8	0.202419	0.195	-0.20242	78	-0.20242
2	0.183156	8	0.183156	0.2	-0.18316	80	-0.18316
2.05	0.165727	8.2	0.165727	0.205	-0.16573	82	-0.16573
2.1	0.149956	8.4	0.149956	0.21	-0.14996	84	-0.14996
2.15	0.135686	8.6	0.135686	0.215	-0.13569	86	-0.13569
2.2	0.122773	8.8	0.122773	0.22	-0.12277	88	-0.12277
2.25	0.11109	9	0.11109	0.225	-0.11109	90	-0.11109
2.3	0.100518	9.2	0.100518	0.23	-0.10052	92	-0.10052
2.35	0.090953	9.4	0.090953	0.235	-0.09095	94	-0.09095
2.4	0.082297	9.6	0.082297	0.24	-0.0823	96	-0.0823
2.45	0.074466	9.8	0.074466	0.245	-0.07447	98	-0.07447
2.5	0.067379	10	0.067379	0.25	-0.06738	100	-0.06738

The plots of the functions are shown in Figures P5-6sa and P5-6sb. We need to show the functions with two time scales in order to see the characteristic shape of the curve. A and C are seen in Figure P5-6sa for at least 5 T_C. B, the upper most curve in Figure P5-6sa, and D, the lowest curve are seen for at least 5 T_C in Figure P5-6sb.

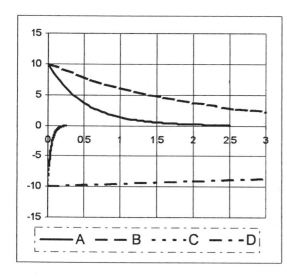

Figure P5-6sa

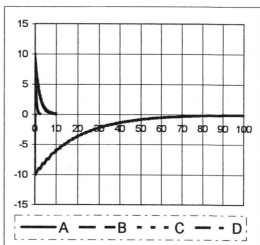

Figure P5-6sb

5-11 Determine the period, frequency, amplitude, time shift, and phase angle of the following sinusoids. Graph the waveforms.

(a) $v_1(t) := 10 \cdot \cos(2000 \cdot \pi \cdot t) + 10 \cdot \sin(2000 \cdot \pi \cdot t)$

(b) $v_2(t) := -30 \cdot \cos(2000 \cdot \pi \cdot t) - 20 \cdot \sin(2000 \cdot \pi \cdot t)$

(c) $v_3(t) := 10 \cdot \cos\left(\dfrac{2 \cdot \pi \cdot t}{10}\right) - 10 \cdot \sin\left(\dfrac{2 \cdot \pi \cdot t}{10}\right)$

(d) $v_4(t) := -20 \cdot \cos(800 \cdot \pi \cdot t) + 30 \cdot \sin(800 \cdot \pi \cdot t)$

Solutions: The general sinusoid is in the form

$$v(t, a, b, f) := a \cdot \cos(2 \cdot \pi \cdot f \cdot t) + b \cdot \sin(2 \cdot \pi \cdot f \cdot t)$$

This can be written in the form

$$v\left(t, V_A, f, \phi\right) := V_A \cdot \cos(2 \cdot \pi \cdot f \cdot t + \phi)$$

where $\quad V_A(a, b) := \sqrt{a^2 + b^2}$

and $\quad \phi(a, b) := \text{atan2}(a, -b)$ (Please check help to remind yourself of the use of the atan2 function.)

So, let's get to the solutions. The frequencies and periods of the waveforms are:

$f_1 := 1000 \cdot \text{Hz}$ $\qquad T_1 := \dfrac{1}{f_1}$ $\qquad T_1 = 1 \times 10^{-3} \, \text{s}$

$f_2 := 1000 \cdot \text{Hz}$ $\qquad T_2 := \dfrac{1}{f_2}$ $\qquad T_2 = 1 \times 10^{-3} \, \text{s}$

$f_3 := .1 \cdot \text{Hz}$ $\qquad T_3 := \dfrac{1}{f_3}$ $\qquad T_3 = 10 \, \text{s}$

$f_4 := 400 \cdot \text{Hz}$ $\qquad T_4 := \dfrac{1}{f_4}$ $\qquad T_4 = 2.5 \times 10^{-3} \, \text{s}$

The amplitudes are phase angles (in degrees) are:

$V_1 := V_A(10, 10)$ $\qquad V_1 = 14.142 \, \text{V}$ $\qquad \phi_1 := \phi(10, 10)$ $\qquad \phi_1 = -0.785$

$V_2 := V_A(-30, -20)$ $\qquad V_2 = 36.056 \, \text{V}$ $\qquad \phi_2 := \phi(-30, -20)$ $\qquad \phi_2 = 2.554$

$V_3 := V_A(10, 10)$ $\qquad V_3 = 14.142 \, \text{V}$ $\qquad \phi_3 := \phi(10, -10)$ $\qquad \phi_3 = 0.785$

$V_4 := V_A(-20, 30)$ $\qquad V_4 = 36.056 \, \text{V}$ $\qquad \phi_4 := \phi(-20, 30)$ $\qquad \phi_4 = -2.159$

Finally, the time shift is given by

$$T_S(\phi, T_0) := \frac{\phi \cdot T_0}{2 \cdot \pi}$$

$$T_{S1} := T_S(\phi_1, T_1) \qquad T_{S1} = 1.25 \times 10^{-4}\, s$$

$$T_{S2} := T_S(\phi_2, T_2) \qquad T_{S2} = -4.064 \times 10^{-4}\, s$$

$$T_{S3} := T_S(\phi_3, T_3) \qquad T_{S3} = -1.25\, s$$

$$T_{S4} := T_S(\phi_4, T_4) \qquad T_{S4} = 8.59 \times 10^{-4}\, s$$

The graphs of the functions follow. We want to plot at least two complete cycles of the waveforms going through t = 0 and with a total of 50 points. To do this, we create the time sequences based on the periods of the waveforms.

$$t_1 := -T_1, -T_1 + \frac{T_1}{25} \, .. \, T_1 \qquad\qquad t_2 := -T_2, -T_2 + \frac{T_2}{25} \, .. \, T_2$$

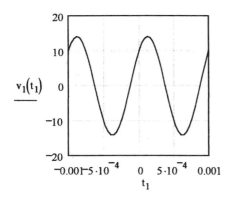

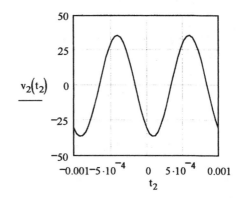

$$t_3 := -T_3, -T_3 + \frac{T_3}{25} \, .. \, T_3 \qquad\qquad t_4 := -T_4, -T_4 + \frac{T_4}{25} \, .. \, T_4$$

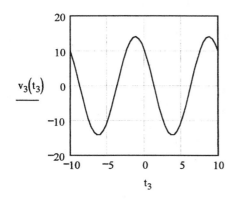

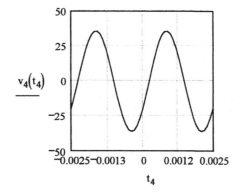

41

5-18 Graph the following waveforms.

    (a) $v_1(t) = 10 [1 - 2e^{-200t} \sin(1000\pi t)]u(t)$

    (b) $v_2(t) = [20 - 10e^{-1000t}]u(t)$

    (c) $v_3(t) = 10 [2 - \sin(1000\pi t)]u(t)$

    (d) $v_4(t) = 10 [4 - 2e^{-1000t}]u(t)$

Solution: We will use MATLAB to do this problem. We want to plot at least 5 time constants (5 T_C) for each plot with an exponential. For (c), since there is no exponential, we want to plot at least two periods. The m-file below addresses these issues.

```
% Problem 5-18 asks for graphs of the following functions:
% (a)  v1(t) = 10 [1 - 2exp(-200t) sin(1000 pi t)]u(t)
%
f1=500;
T1=1/f1;
TC1=1/200;
t1=0:T1/25:5*TC1;
v1=10*(1-2*exp(-200*t1).*sin(1000*pi*t1));
%
% (b)  v2(t) = [20 - 10exp(-1000t)]u(t)
%
TC2=1/1000;
t2=0:TC2/10:5*TC2;
v2=20-10*exp(-1000*t2);
%
% (c)  v3(t) = 10 [2 -  sin(1000 pi t)]u(t)
%
f3=500;
T3=1/f3;
t3=0:T3/25:2*T3;
v3=10*(2-sin(1000*pi*t3));
%
% (d)  v4(t) = 10 [4 - 2exp(-1000t)]u(t)
%
TC4=1/1000;
t4=0:TC4/10:5*TC4;
v4=10*(4-2*exp(-1000*t4));
%
% Finally, create a plot with all four functions contained in sub-plots.
%
subplot(2,2,1),plot(t1,v1),title('Problem 5-18 (a)')
xlabel('time in seconds')
ylabel('v1(t) (volts)')
grid
subplot(2,2,2),plot(t2,v2),title('Problem 5-18 (b)')
xlabel('time in seconds')
ylabel('v2(t) (volts)')
grid
subplot(2,2,3),plot(t3,v3),title('Problem 5-18 (c)')
xlabel('time in seconds')
ylabel('v3(t) (volts)')
grid
subplot(2,2,4),plot(t4,v4),title('Problem 5-18 (4)')
xlabel('time in seconds')
ylabel('v4(t) (volts)')
grid
```

The resulting graphs follow:

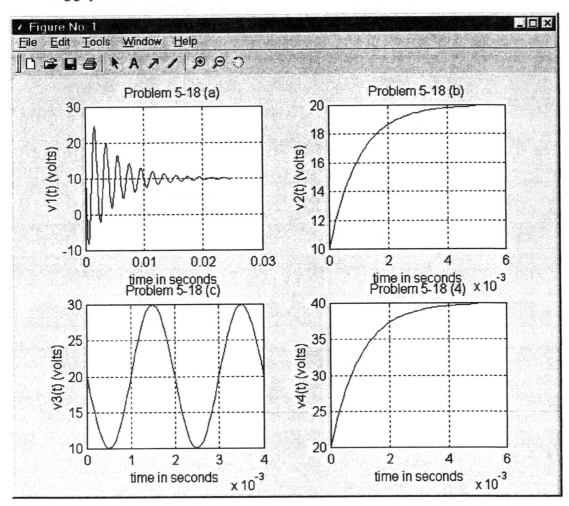

Figure P5-18s

5-21 A waveform is known to be of the form $v(t) = V_A e^{-\alpha t} \sin \beta t$. The waveform periodically passes through zero every 2.5 ms. At $t = 1$ ms its value is 3.5 V and at $t = 2$ ms it is 0.8 V. Find the values of the parameters V_A, α, and β, and then graph the waveform.

Solution: We know that the waveform has a half-period of 2.5 ms. We can then find the period T_0, the frequency f and β from.

$$T_0 := 2 \cdot 2.5 \cdot 10^{-3} \cdot s$$

$$f := \frac{1}{T_0} \qquad\qquad f = 200\,\mathrm{Hz}$$

$$\beta := 2 \cdot \pi \cdot f \qquad\qquad \beta = 1.257 \times 10^3\,\mathrm{s}^{-1}$$

We are now left with two unknowns, α and V_A. These can be found from the other two pieces of information. We set up a solve block to solve two equations. The guesses are

$$V_A := 5 \qquad\qquad \alpha := 500$$

Given

$$V_A \cdot e^{-\alpha \cdot 10^{-3}} \cdot \sin\left(\beta \cdot 10^{-3} \cdot s\right) = 3.5 \qquad\qquad V_A \cdot e^{-\alpha \cdot 2 \cdot 10^{-3}} \cdot \sin\left(\beta \cdot 2 \cdot 10^{-3} \cdot s\right) = 0.8$$

$$\begin{pmatrix} V_A \\ \alpha \end{pmatrix} := \mathrm{Find}(V_A, \alpha) \qquad\qquad \begin{pmatrix} V_A \\ \alpha \end{pmatrix} = \begin{pmatrix} 9.951 \\ 994.695 \end{pmatrix}$$

The waveform then has the final form

$$v(t) := V_A \cdot e^{-\alpha \cdot t} \cdot \sin(\beta \cdot t) \quad \mathrm{V}$$

To graph the function, we set of a series of t out to at lease five time constants, $1/\alpha$.

$$t := 0, \frac{5}{100 \cdot \alpha} \, .. \, \frac{10}{\alpha}$$

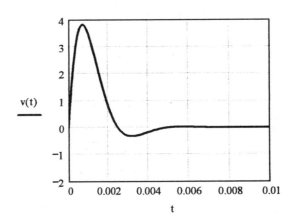

5-42 (A) ANALOG-TO-DIGITAL CONVERSION (A)

Figure P5-42 shows a circuit diagram of an analog-to-digital converter based on a voltage divider and OP AMPs operating without feedback. Each of the OP AMPs operates as a comparator whose output is $V_{OH} = 10$ V when $v_P > v_N$ and $V_{OL} = 0$ V when $v_P < v_N$. The input signal $v_S(t)$ is applied to all of the noninverting inputs simultaneously. The voltages applied to the inverting inputs come from the four-resistor voltage divider shown. For an input voltage of $v_S(t) = 25e^{-2000t}$ V, express the output voltages $v_A(t)$, $v_B(t)$, $v_C(t)$, and $v_D(t)$ in terms of step functions.

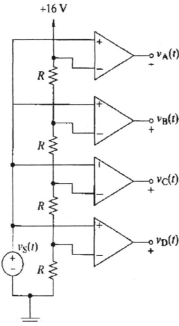

Figure P5-42

Solution: The voltage divider applies voltages of 16 V, 12 V, 8 V and 4 V to the inverting terminals of OP AMPS A, B, C and D respectively. These are then comparied to $v_S(t)$ applied to the non-inverting terminals. The output voltages can be written as:

$$v_S(t) := 25 \cdot e^{-2000 \cdot t} \cdot volt \qquad V_{OH} := 10 \cdot volt \qquad V_{OL} := 0 \cdot volt$$

$$V_A(t) := \begin{vmatrix} V_{OH} & \text{if } v_S(t) > 16 \\ V_{OL} & \text{otherwise} \end{vmatrix} \qquad V_B(t) := \begin{vmatrix} V_{OH} & \text{if } v_S(t) > 12 \\ V_{OL} & \text{otherwise} \end{vmatrix}$$

$$V_C(t) := \begin{vmatrix} V_{OH} & \text{if } v_S(t) > 8 \\ V_{OL} & \text{otherwise} \end{vmatrix} \qquad V_D(t) := \begin{vmatrix} V_{OH} & \text{if } v_S(t) > 4 \\ V_{OL} & \text{otherwise} \end{vmatrix}$$

We start the time sequence at 0 and create 100 points over the time interval to 3 T_C, where

$$T_C := \frac{1}{2000} \cdot s^{-1}$$

$$t := 0, \frac{3T_C}{100} \, .. \, 2T_C$$

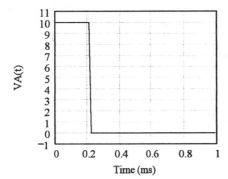

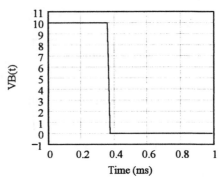

$$t := 0, \frac{3T_C}{100} \, .. \, 3T_C$$

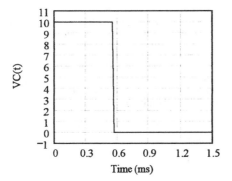

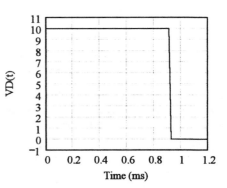

6-6 The voltage across a 0.5-μF capacitor is shown in Figure P6-6. Derive expressions for, and sketch, $i_C(t)$, $p_C(t)$, and $w_C(t)$. Is the capacitor absorbing power, delivering power, or both?

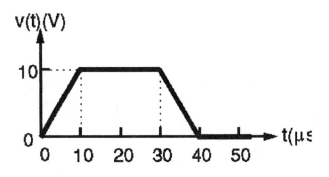

Figure P6-6

Solution: First we must write the expression for the voltage.

$$v_C(t) = \frac{10}{10 \times 10^{-6}} t \left[u(t) - u(t - 10 \times 10^{-6})\right] + 10 \left[u(t - 10 \times 10^{-6}) - u(t - 30 \times 10^{-6})\right]$$

$$- \frac{10}{10 \times 10^{-6}} t \left[u(t - 30 \times 10^{-6}) - u(t - 40 \times 10^{-6})\right]$$

For the current through the capacitor, we have

$$i_C(t) = C \frac{dv_C(t)}{dt} = \frac{5 \times 10^{-6}}{10 \times 10^{-6}} \left[u(t) - u(t - 10 \times 10^{-6})\right]$$

$$- \frac{5 \times 10^{-6}}{10 \times 10^{-6}} \left[u(t - 30 \times 10^{-6}) - u(t - 40 \times 10^{-6})\right]$$

The expressions for $p_C(t)$ and $w_C(t)$ are

$$p_C(t) = v_C(t) i_C(t)$$

$$w_C(t) = \int v_C(t) i_C(t) \, dt = \frac{1}{2} C v_C^2(t)$$

We now use Excel to do the calculations and graphs for $p(t)$, $w(t)$, and $i_C(t)$. The graphs follow. From the graph of p(t), we see that the capacitor absorbs power when the values are positive, and delivers power when the values are negative.

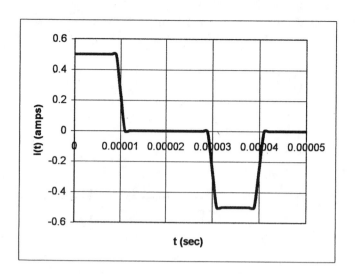

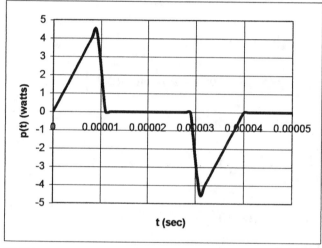

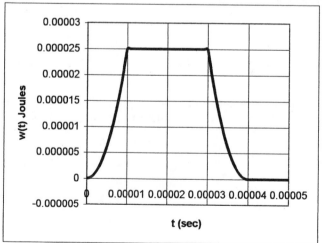

6-9 For t > 0, the current through a 500-mH inductor is $i_L(t) = 10\,e^{-2000t}\sin 1000t$ mA. Derive expressions for, and sketch, $v_L(t)$, $p_L(t)$, and $w_L(t)$. Is the inductor absorbing power, delivering power, or both?

Solution: For the given current, the voltage across the inductor is

$$v_L(t) = L\frac{di(t)}{dt} = 500\times10^{-3}\times10\left(-2000\,e^{-2000t}\sin 1000t + 1000\,e^{-2000t}\cos 1000t\right)mV$$

$$= 5000\,e^{-2000t}\left(-2\sin 1000t + \cos 1000t\right)mV$$

The power and energy are given by

$$p_L(t) = v_L(t)\,i_L(t)$$

$$w_L(t) = \frac{1}{2}i_L^2(t)\,L$$

We now use MATLAB to do the calculations and graphing. The m-file and graphs follow. From the graph of p(t), we see that the inductor absorbs power when the values are positive, and delivers power when the values are negative.

```
% Set up the t array to be able to plot 5 time constants of the function.
%
t=0:5/200000:5/2000;
%
% Define the inductance, current, voltage, power and energy functions.
%
L=.5;
iL=10*10^(-3)*exp(-2000*t).*sin(1000*t);
vL=5000*10^(-3)*exp(-2000*t).*(-2*sin(1000*t)+cos(1000*t));
pL=vL.*iL;
wL=1/2*(iL.^2)*L;
%
% Plot the functions
%
subplot(2,2,1),plot(t,iL),title('Problem 6-9 (a)')
xlabel('time in seconds')
ylabel('iL(t) (amps)')
grid
subplot(2,2,2),plot(t,vL),title('Problem 6-9 (b)')
xlabel('time in seconds')
ylabel('vL(t) (volts)')
grid
subplot(2,2,3),plot(t,pL),title('Problem 6-9 (c)')
xlabel('time in seconds')
ylabel('pL(t) (watts)')
grid
subplot(2,2,4),plot(t,wL),title('Problem 6-9 (d)')
xlabel('time in seconds')
ylabel('wL(t) (joules)')
grid
```

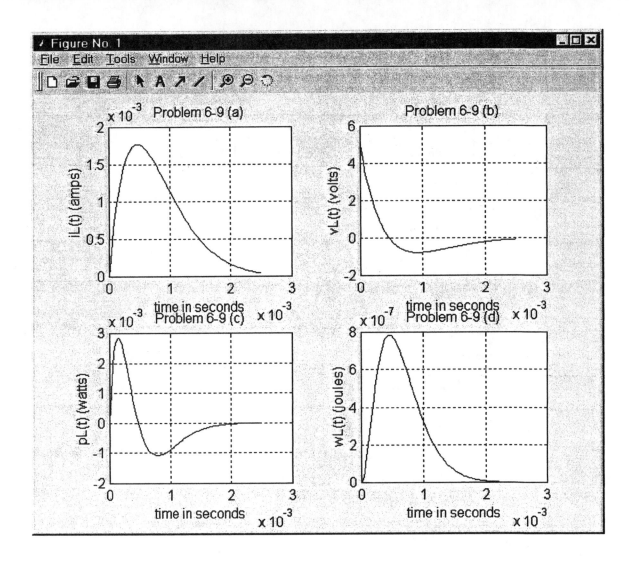

6-12 The current through a 25-mH inductor is shown in Figure P6-7. Derive expressions for, and sketch, $v_L(t)$, $p_L(t)$, and $w_L(t)$.

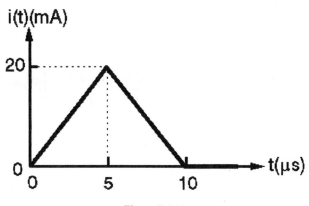

i(t)(mA)

Figure P6-12

Solution: First we need to write down the expression for the current from the Figure P6-12.

$$i_L(t) = \frac{20 \times 10^{-3}}{5 \times 10^{-6}} t \left[u(t) - u(t - 5 \times 10^{-6})\right] - \frac{20 \times 10^{-3}}{5 \times 10^{-6}} t \left[u(t - 5 \times 10^{-6}) - u(t - 10 \times 10^{-6})\right]$$

$$= 4000 t \left[u(t) - 2u(t - 5 \times 10^{-6}) + u(t - 10 \times 10^{-6})\right]$$

The voltage across the inductor is then given by

$$v_L(t) = L\frac{di_L(t)}{dt} = .025 \times 4000 \left[u(t) - 2u(t - 5 \times 10^{-6}) + u(t - 10 \times 10^{-6})\right]$$

$$= 100 \left[u(t) - 2u(t - 5 \times 10^{-6}) + u(t - 10 \times 10^{-6})\right]$$

Finally, $p_L(t)$ and $w_L(t)$ are given by

$$p_L(t) = v_L(t) i_L(t)$$

$$w_L(t) = \frac{1}{2} i_L^2(t) L$$

We use Excel to calculate and plot the functions. The graphs follow. From the graph of p(t), we see that the inductor absorbs power when the values are positive, and delivers power when the values are negative.

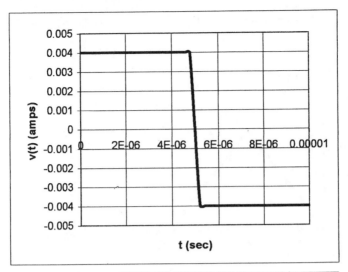

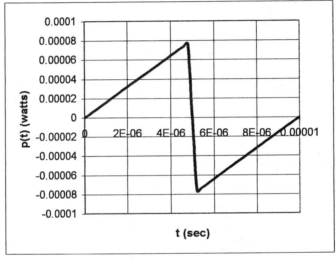

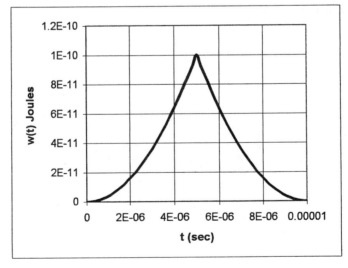

6-40 Find the equivalent capacitance between terminals A and B in Figure P6-40.

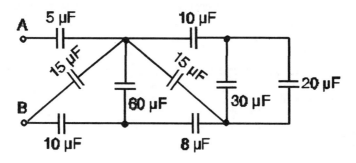

Figure P6-40

Solution: We can work from the right using parallel and series combinations:

$$C_{eq} := \left[\left[\left[\left[\left[\left[(20+30)^{-1} + 10^{-1} \right]^{-1} + 15 \right]^{-1} + 8^{-1} \right]^{-1} + 60 \right]^{-1} + 10^{-1} \right]^{-1} + 15 \right]^{-1} + 5^{-1} \right]^{-1} \cdot \mu F$$

$C_{eq} = 4.128 \, \mu F$

6-42 A capacitor bank is required that can be charged to 5 kV and store at least 250 J of energy. Design a series/parallel combination that meets the voltage and energy requirements using 20 μF capacitors each rated at 1.5 kV max.

Solution: First we define the input values:

$$V_{charge} := 5000 \cdot volt \qquad W_{stored} := 250 \cdot joule \qquad C := 20 \cdot \mu F \qquad V_{rated} := 1500 \cdot volt$$

Now calculate the total capacitance needed.

$$C_{total} := 2 \cdot \frac{W_{stored}}{V_{charge}^2} \qquad\qquad C_{total} = 20 \, \mu F$$

We can connect a number of capacitors in series to achieve the 1.5 kV rating. To do this, we need at least

$$\frac{V_{charge}}{V_{rated}} = 3.333$$

capacitors. If we choose four, the total capacitance will be

$$C_4 := \frac{C}{4} \qquad\qquad C_4 = 5 \times 10^{-6} \, F$$

If we now take the four in series as a single equivalent capacitor, in put it in parallel with three other such equivalent capacitors, we will have the needed 20 μF total capacitance. We therefore need a total of 16 - 20 μF capacitors.

6-46 (A) CAPACITIVE DISCHARGE PULSER

A capacitor bank for a large pulse generator consists of 11 capacitor strings connected in parallel. Each string consists of 16 1.5-mF capacitors connected in series. The purpose of this problem is to calculate important characteristics of the pulser.
(a) What is the total equivalent capacitance of the bank?
(b) If each capacitor in a series string is charged to 300 V, what is the total energy stored in the bank:
(c) In the discharge mode, the voltage across the capacitor bank is $v(t) = 4.8[e^{-500t}]u(t)$ kV. What is the peak power delivered by the capacitor bank?
(d) For practical purposes the capacitor bank is completely discharged after about five time constants. What is the average power delivered during that interval?

Solution: The MATLAB m-file follows and is fairly straightforward. The only explanation needed is for the "peak power" and the "average power". The peak power is where the voltage and current functions are maximum. They are both maximum at t = 0. Since $i(t) = C \, dv/dt$, $i(t) = C \times 4800 \times (-500) [e^{-500t}]u(t)$, and the peak power is $C \times 4800 \times (-500)$. The average power is the integral of the power function over the specified time period. The time constant of the discharge function is 1/500.

```
% (a) Calculate the total capacitance of the bank.
C=1.5*10^(-3);
Cstring=C/16;
Cbank=11*Cstring
% (b) Calculate the total energy in the bank.
Vc=300;
Vstring=16*Vc;
Wbank=0.5*Cbank*Vstring^2
% (c) Calculate the peak power delivered by the capacitor bank.
% This is the power at t=0.
v0=4800
i0=Cbank*v0*500
p0=v0*i0
% (d) Calculate the average power delivered during 5 time constants.
TC=1/500;
syms p i v t
v=4800*exp(-500*t);
i=-Cbank*4800*500*exp(-500*t);
p=i*v;
paverage=int(p,t,0,TC)
```

The result of running the m-file follows:

Cbank = 0.00103125000000
Wbank = 11880
v0 = 4800
i0 = 2475
p0 = 11880000
paverage = 11880*exp(-1)^2-11880

Finally, since the int function gives us an expression, we simply copy and paste the expression into the command line to evaluate it. The result is

Ans = -1.027221683514 904e+004

7-7 The switch in each circuit in Figure P7-7 has been in position A for a long time and is moved to position B at $t = 0$. For each circuit write an expression for the state variable for $t \geq 0$ and sketch its waveform.

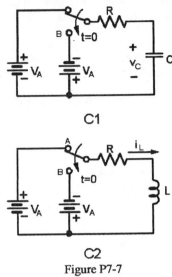

C1

C2

Figure P7-7

Solution: We will use Orcad Capture for this solution. The schematic for the RC circuit is shown below. The VPULSE (V2) is set up for a V1 of 5 and a V2 of –5 and an ideal pulse of 0 rise time and 0 fall time.

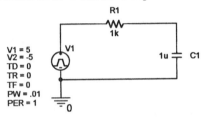

Figure P7-7sa

Since the time constant for this circuit is 1 ms, we will set the transient response (Analysis - Setup - Transient) for a length of 5 ms. The resulting transient response is shown in Figure P7-7sb.

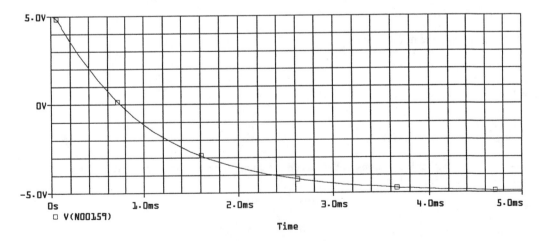

Figure P7-7sb

The expression for $v_C(t)$ is

$$v_C(t) = FV + (IV - FV)e^{-t/T_c} = -V_A + 2V_A e^{-t/RC} \text{ V}, \text{ t} \geq 0$$

The schematic for the RL circuit is shown below.

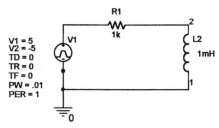

Figure P7-7sc

Since the time constant for this circuit is 1 μs, we will set the transient response (Analysis - Setup - Transient) for a length of 5 μs. The resulting transient response is shown in Figure P7-7sd.

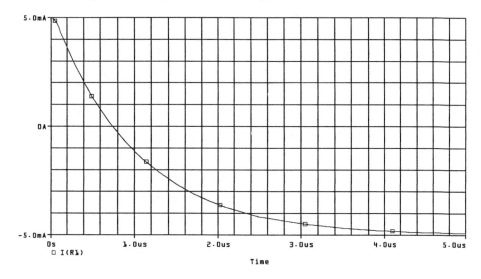

Figure P7-7sd

The expression for $i_L(t)$ is

$$i_L(t) = FV + (IV - FV)e^{-t/T_c} = -\frac{V_A}{R} + 2\frac{V_A}{R}e^{-tL/R} \text{ A}, \text{ t} \geq 0$$

7-13 The switch in Figure P7-13 has been open for a long time and is closed at $t = 0$. Solve for $v_C(t)$ for $t \geq 0$ when the input is $v_S(t) = 15$ V. Identify the forced and natural components, and plot its waveform.

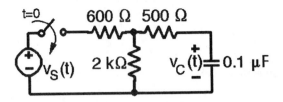

Figure P7-13

Solution: By inspection, the initial value and the final value of the voltage across the capacitor are 0 V and 2000/2600 % 15 = 11.538 V respectively. The time constant is $R_T C$, where R_T is given by

$$R_T = 500 + \frac{600 \times 2000}{600 + 2000} = 961.5 \ \Omega$$

so that $R_T C = 96.15$ μs.

The expression for $v_C(t)$ is then

$$v_C(t) = FV + (IV - FV)e^{-t/T_C} = 11.538 - 11.538 \ e^{-t/R_T C} \quad V$$

where the forced and natural responses are given by

$$v_{CF}(t) = 11.538 \ V$$
$$v_{CN}(t) = -11.538 \ e^{-t/R_T C} \quad V$$

We use Excel to plot the function $v_C(t)$.

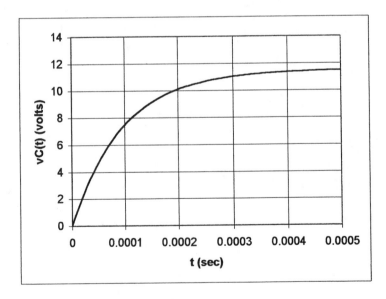

Figure 7-13s

7-14 The input in Figure P7-13 is $v_S(t) = 15$ V. The switch has been open for a long time and is closed at $t = 0$. The switch is subsequently reopened at $t = 200$ μs. Solve for $v_C(t)$ for $t \geq 0$ and plots its waveform.

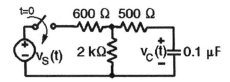

Figure P7-13

Solution: By inspection, the initial value and the final value of the voltage across the capacitor if the switch were to remain closed are 0 V and 2000/2600 % 15 = 11.538 V respectively. The expression for the first 200 μs is then

$$v_C(t) = FV + (IV - FV)e^{-t/T_C} = 11.538 - 11.538\,e^{-t/R_{T1}C} \quad V$$

where R_{T1} is given by

$$R_{T1} = 500 + \frac{600 \times 2000}{600 + 2000} = 961.5\,\Omega$$

so that $R_{T1}C = 96.15$ μs. At 200 μs, v_C is

$$v_C(0.0002) = 11.538 - 11.538\,e^{-0.0002/0.00009615} = 10.10\,V$$

Now the capacitor will begin discharging through the series combination of the 500 Ω and 2 kΩ resistors ($R_{T2} = 2500\,\Omega$). The time constant is then $R_{T2}C = 0.25$ ms. The expression for $v_C(t)$ for $t \geq 200$ μs is then

$$v_C(t) = FV + (IV - FV)e^{-(t-.0002)/T_C} = -10.10\,e^{-(t-0.0002)/R_{T2}C} \quad V$$

We created an m-file in MATLAB to calculate the above values and to plot the entire function.

```
% Define the circuit components.
%
R1=600;
R2=2000;
R3=500;
C=0.1*10^(-6);
VS=15;
%
% Calculate the time constant and Thevenin Voltage for t < .200 ms.
%
RT1=R3+(R1*R2/(R1+R2))
TC1=RT1*C
VT=R2/(R2+R1)*VS
%
% Calculate the time constant for t > .200 ms and the voltage at t = .200 ms.
%
RT2=R2+R3
TC2=RT2*C
t200=200*10^(-6);
V200=VT-VT*exp(-t200/TC1)
%
```

```
% Set up the t array to be able to plot 5 time constants of the function.
%
t=0:0.00001:0.00125;
t1=0:0.00001:0.0002;
t2=0.00021:0.00001:0.00125;
%
% Define the function for the two time intervals.
%
vC=VT-VT*exp(-1/TC1*t1);
vC((length(t1)+1):length(t))=V200*exp(-1/TC2*(t2-t200));
%
% Plot the function.
%
plot(t,vC)
xlabel('sec')
ylabel('vC(t) (Volts)')
%
```

The values and resulting graph follow:

RT1 = 9.615384615384616e+002
TC1 = 9.615384615384615e-005
VT = 11.53846153846154
RT2 = 2500
TC2 = 2.500000000000000e-004
V200 = 10.09695909001636

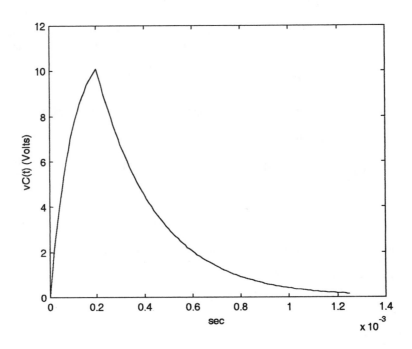

7-36 Traditional, 8-6 Laplace Early. The switch in Figure P7-36 has been open for a long time and is closed at t = 0. The circuit parameters are $L = 0.5$ H, $C = 25$ nF, $R_1 = 5$ kΩ, $R_2 = 12$ kΩ, and $V_A = 10$ V.
(a) Find the initial values of v_C and i_L at $t = 0$ and the final values of v_C and i_L as $t \to \infty$.
(b) Find the differential equation for $v_C(t)$ and the circuit characteristic equation.
(c) Find $v_C(t)$ and $i_L(t)$ for $t \geq 0$, identify the forced and natural components of the responses, and plot $v_C(t)$ and $i_L(t)$. Is the circuit overdamped or underdamped?

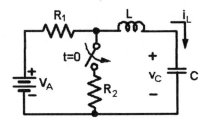

Figure P7-36

Solution:
(a) The initial values of v_C and i_L are, by inspection, 10 V and 0 A respectively. The final values of v_C and i_L are 7.05 V and 0 A.
(b) At t = 0 the circuit can be redrawn as a series RLC circuit with the series resistance being $R_T = R_1 \parallel R_2$ and the Thévenin voltage equal to 7.05 V. The circuit differential and characteristic equations are then:

$$LC\frac{d^2}{dt^2}v_C(t) + R_T C\frac{d}{dt}v_C(t) + v_C(t) = 7.05\,u(t)$$

$$LC\,s^2 + R_T C\,s + 1 = 0$$

(c) This exercise concentrates on the transient analysis of the circuit using the initial condition on the capacitor. Mathcad or MATLAB can be used to solve for $v_C(t)$ and $i_L(t)$ respectively[1]. The resulting equations are:

$$v_C(t) = e^{-3529t}\left(2.941\cos(8213t) + 1.263\sin(8213t)\right)V$$

$$i_C(t) = C\frac{d}{dt}v_C(t)\,A$$

We now simulate the circuit in Electronics Workbench. The circuit schematic is shown in Figure P7-36sa.

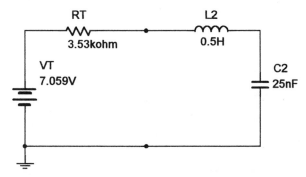

Figure P7-36sa

---

[1] See problem 7-37 for a detailed Mathcad solution of this type of problem.

The initial condition for the capacitor is set in the values dialog box for the capacitor – see Figure P7-36sb.

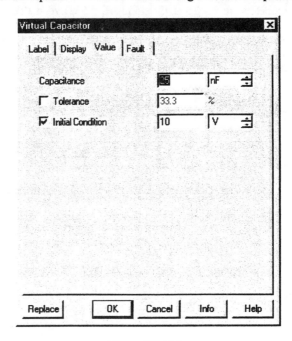

Figure P7-36sb

The transient analysis is set up by choosing Simulate - Analyses - Transient Analysis menu. The dialog box shown in Figure P7-36sc appears. Under the Analysis Parameters tab, "User-defined" is selected under "Initial conditions". We set the Analysis start and stop times using the parameters of the v_C (t) solution, and set the minimum number of time points to a reasonable value to get good resolution.

Figure P7-36sc

62

The Output Variables tab brings up another dialog box shown in Figure P7-36sd. Variables can be selected from the left for display on the right. We choose the voltage node 4 to display $v_C(t)$ first and then simulate.

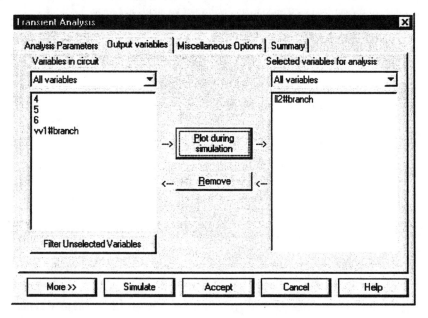

Figure P7-36sd

Figure 7-36se is the result. Then we select II2#branch to display $i_L(t)$. The simulated result is shown in Figure 7-36sf. The reason we do on simulation at a time is that the scales are very different for each value. If you tried to do both, the current would be a straight line since the values are in the μA range. Note that the responses are typical of the underdamped system that we are analyzing. The circuit has a ζ of 0.395.

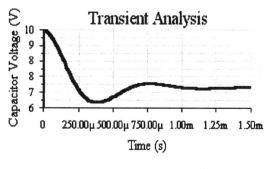

Figure P7-36se

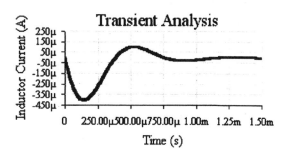

Figure P7-36sf

7-37 Traditional, 8-7 Laplace Early. The switch in Figure P7-36 has been closed for a long time and is opened at $t = 0$. The circuit parameters are $L = 0.5$ H, $C = 0.4$ μF, $R_1 = 1$ kΩ, $R_2 = 2$ kΩ, and $V_A = 15$ V.

(a) Find the initial values of v_C and i_L at $t = 0$ and the final values of v_C and i_L as $t \to \infty$.

(b) Find the differential equation for $v_C(t)$ and the circuit characteristic equation.

(c) Find $v_C(t)$ and $i_L(t)$ for $t = 0$, identify the forced and natural components of the responses, and plot $v_C(t)$ and $i_L(t)$. Is the circuit overdamped or underdamped?

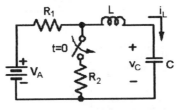

Figure P7-36

Solution: First we need a few definitions for circuit components.

$$R_1 := 1000 \qquad R_2 := 2000 \qquad L := .5 \qquad C := 0.4 \cdot 10^{-6} \qquad V_A := 15$$

(a) The initial values of v_C and i_L at $t = 0$ are, by inspection, 10 V and 0 A respectively. As $t \to \infty$, v_C and i_L will go to V_A and 0 A respectively. After the switch is opened, the Thevenin Resistance for the circuit is:

$$R_T := R_1$$

(b) The differential equation for $v_C(t)$ and the circuit characteristic equation are given by:

$$L \cdot C \cdot \frac{d^2}{dt^2} v_C(t) + R_T \cdot C \cdot \frac{d}{dt} v_C(t) + v_C(t) = V_A$$

$$L \cdot C \cdot s^2 + R_T \cdot C \cdot s + 1 = 0$$

(c) The solution to the characteristic equation can be accomplished using the polyroots function.

$$\begin{pmatrix} \alpha_1 \\ \alpha_2 \end{pmatrix} := \text{polyroots}\left(\begin{pmatrix} 1 & R_T \cdot C & L \cdot C \end{pmatrix}^T\right) \qquad \begin{pmatrix} \alpha_1 \\ \alpha_2 \end{pmatrix} = \begin{pmatrix} -1 \times 10^3 - 2i \times 10^3 \\ -1 \times 10^3 + 2i \times 10^3 \end{pmatrix}$$

Since the roots are complex the circuit is underdamped. We define

$$\alpha := -\text{Re}(\alpha_1) \qquad \beta := \text{Im}(\alpha_2)$$

so that the general solutions for $v_C(t)$ and $i_L(t)$ are then

$$v_C(t) = e^{-\alpha \cdot t} \cdot \left(K_1 \cdot \cos(\beta \cdot t) + K_2 \cdot \sin(\beta \cdot t)\right) + V_A$$

$$i_L(t) = C \cdot \left(\frac{d}{dt} v_C(t)\right) = C \cdot e^{-\alpha \cdot t} \cdot \left[-\alpha \cdot \left(K_1 \cdot \cos(\beta \cdot t) + K_2 \cdot \sin(\beta \cdot t)\right) + \beta \cdot \left(-K_1 \cdot \sin(\beta \cdot t) + K_2 \cdot \cos(\beta \cdot t)\right)\right]$$

64

where V_A is the forced response for $v_C(t)$ and the rest of $v_C(t)$ is the natural response. $i_L(t)$ is all natural response. We now evaluate K_1 and K_2 from the initial conditions. $v_C(0) = 10$ implies

$$K_1 := -5$$

$$\frac{d}{dt}v_C(t) = 0 \quad \text{implies}$$

$$-\alpha \cdot K_1 + \beta \cdot K_2 = 0 \quad \text{or} \quad K_2 := \frac{\alpha \cdot K_1}{\beta} \quad K_2 = -2.5$$

We now rewrite the expressions for $v_C(t)$ and $i_L(t)$ for plotting purposes.

$$v_C(t) := e^{-\alpha \cdot t} \cdot \left(K_1 \cdot \cos(\beta \cdot t) + K_2 \cdot \sin(\beta \cdot t) \right) + V_A$$

$$i_L(t) := C \cdot e^{-\alpha \cdot t} \cdot \left[-\alpha \cdot \left(K_1 \cdot \cos(\beta \cdot t) + K_2 \cdot \sin(\beta \cdot t) \right) + \beta \cdot \left(-K_1 \cdot \sin(\beta \cdot t) + K_2 \cdot \cos(\beta \cdot t) \right) \right]$$

$$t := 0, 10^{-5} .. 5 \cdot 10^{-3}$$

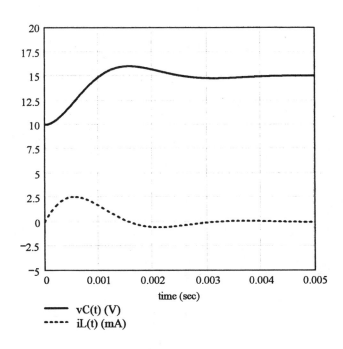

65

7-42 Traditional, 8-12 Laplace Early. The switch in the circuit shown in Figure P7-42 has been in position A for a long time. At $t = 0$ it is moved to position B. The circuit parameters are $R_1 = 1$ kΩ, $R_2 = 4$ kΩ, $L = 0.625$ H, $C = 6.25$ nF, and $V_A = 15$ V. Find $v_C(t)$ and $i_L(t)$ for $t > 0$, identify the forced and natural components of the responses, and plot $v_C(t)$ and $i_L(t)$. Is the circuit overdamped or underdamped?

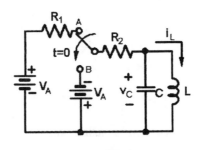

Figure P7-42

Solution: First, we can replace V_A and R_2 by their Norton equivalent, $i_N = V_A/R_2$ and $G_N = 1/R_2$. We then have a parallel RLC circuit and we can write down the differential equation for $i_L(t)$.

$$LC\frac{d^2}{dt^2}i_L(t) + G_N L\frac{d}{dt}i_L(t) + i_L(t) = i_N$$

The characteristic equation is then

$$LC\,s^2 + G_N L\,s + 1 = 0$$

We now create an MATLAB m-file to find the roots to this equation and plot $v_C(t)$ and $i_L(t)$.

```
% First define the circuit elements.
R1=1000;
R2=4000;
L=0.625;
C=6.25*10^(-9);
VA=15;
% The initial conditions are:
iL0=VA/(R1+R2)
vC0=0
% The final values are:
vCinf=0
iLinf=-VA/R2
% The Norton equivalent values for the source are:
iN=-VA/R2
GN=1/R2
% The coefficient vector for the characteristic polynomial is:
Coef=[L*C,GN*L,1];
% The roots of this equation are found using
alpha=-roots(Coef)
% Now we need to solve for the unknown coefficients using initial conditions.
syms K1 K2
i0=K1+K2+iN-iL0;
di0=alpha(1)*K1+alpha(2)*K2;
[K1,K2]=solve(i0,di0);
k1=double(K1)
k2=double(K2)
```

66

```
% We can finally write down the expressions for iL(t) and
%  vC(t) = L di(t)/dt:
t=0:1/alpha(2)/100:5/alpha(2);
iL=k1*exp(-alpha(1)*t)+k2*exp(-alpha(2)*t)+iN;
vC=-L*(k1*alpha(1)*exp(-alpha(1)*t)+k2*alpha(2)*exp(-alpha(2)*t));
subplot(2,1,1),plot(t,iL,'-')
xlabel('time (sec)')
ylabel('iL(t) (amps)')
subplot(2,1,2),plot(t,vC,'-')
xlabel('time (sec)')
ylabel('vC(t) (volts)')
%
```

The displayed results of the m-file and graphs follow:

iL0 = 0.00300000000000
vC0 = 0
vCinf = 0
iLinf = -0.00375000000000
iN = -0.00375000000000
GN = 2.500000000000000e-004
alpha = 1.0e+004 *
  3.20000000000000
  0.80000000000000
k1 = -0.00225000000000
k2 = 0.00900000000000

The equations for $v_C(t)$ and $i_L(t)$ for $t > 0$ are written in the m-file. This is an overdamped system.

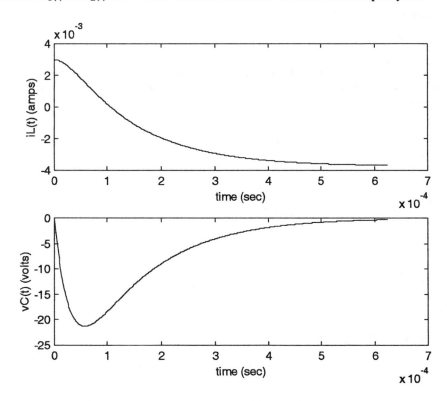

7-63 Traditional, 8-32 Laplace Early. (A) EXPERIMENTAL SECOND-ORDER RESPONSE

Figure P7-63 shows an oscilloscope display of the voltage across the resistor in a series RLC circuit.
(a) Estimate the values of α and β of the damped sine signal.
(b) Use the values of α and β from (a) to write the characteristic equation of the circuit.
(c) The resistor is known to be 2.2 kΩ. Use the characteristic equation from (b) to determine the values of L and C.
(d) If the display shows the zero-input response of the circuit, what are the values of the initial conditions $v_C(0)$ and $i_L(0)$.

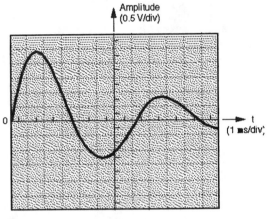

Figure P7-63

(a) Since the waveform is 0 V at t = 0, is of the form

$$v(t) = V_A \cdot e^{-\alpha \cdot t} \cdot \sin(\beta \cdot t)$$

We have satisfied v(0) = 0 with the choice of the sine function, however, we still have 3 unknowns. We therefor need three equations. We pick three points off of the graph to generate the three equations, put them in a solve block with some educated initial guesses.

$$\beta := \frac{2 \cdot \pi}{7 \cdot 10^{-6}} \qquad \beta = 8.976 \times 10^5 \qquad \alpha := \frac{1}{8 \cdot 10^{-6}} \qquad \alpha = 1.25 \times 10^5 \qquad V_A := 2.5$$

Given

$$V_A \cdot e^{-\alpha \cdot \left(1.5 \cdot 10^{-6}\right)} \cdot \sin\left(\beta \cdot 1.5 \cdot 10^{-6}\right) = 2.5$$

$$V_A \cdot e^{-\alpha \cdot \left(11 \cdot 10^{-6}\right)} \cdot \sin\left(\beta \cdot 11 \cdot 10^{-6}\right) = 0$$

$$V_A \cdot e^{-\alpha \cdot \left(10 \cdot 10^{-6}\right)} \cdot \sin\left(\beta \cdot 10 \cdot 10^{-6}\right) = .5$$

$$\begin{pmatrix} V_A \\ \alpha \\ \beta \end{pmatrix} := \text{Find}(V_A, \alpha, \beta) \qquad\qquad \begin{pmatrix} V_A \\ \alpha \\ \beta \end{pmatrix} = \begin{pmatrix} 3.319 \\ 1.613 \times 10^5 \\ 8.568 \times 10^5 \end{pmatrix}$$

68

(b) The characteristic equation of the circuit is then

$$(s + \alpha + j \cdot \beta) \cdot (s + \alpha - j \cdot \beta) \quad \text{evaluation over the complex plane yields} \quad s^2 + 2 \cdot s \cdot \alpha + \alpha^2 + \beta^2$$

Substituting the values for α and β and symbolically evaluating gives:

$$s^2 + 2 \cdot s \cdot \left(1.613 \times 10^5\right) + \left(1.613 \times 10^5\right)^2 + \left(1.613 \times 10^5\right)^2$$

evaluation over the complex plane yields

$$s^2 + 322600.000 \cdot s + 52035380000.000000$$

(c) For a series RLC circuit the characteristic equation is

$$s^2 + \frac{R}{L} \cdot s + \frac{1}{L \cdot C} = s^2 + 2 \cdot s \cdot \alpha + \alpha^2 + \beta^2$$

so that for $R := 2200$ and guesses of $L := .001$ $C := 10^{-9}$

Given

$$\frac{R}{L} = 2 \cdot \alpha \qquad\qquad L \cdot C = \frac{1}{\left(\alpha^2 + \beta^2\right)}$$

$$\begin{pmatrix} L \\ C \end{pmatrix} := \text{Find}(L, C) \qquad\qquad \begin{pmatrix} L \\ C \end{pmatrix} = \begin{pmatrix} 6.821 \times 10^{-3} \\ 1.929 \times 10^{-10} \end{pmatrix}$$

[Note: If you write the equation for LC as:

$$\frac{1}{L \cdot C} = \alpha^2 + \beta^2$$

the solver will not converge. Sometimes one has to rewrite the equations to get the solver to converge.]

(d) Since we already have the initial condition for v(t), v(0) = 0, the initial current can be found from

$$i_L(t) = C \cdot \frac{d}{dt} v(t) = C \cdot \frac{d}{dt} \left(V_A \cdot e^{-\alpha \cdot t} \cdot \sin(\beta \cdot t) \right) = C \cdot V_A \cdot e^{-\alpha \cdot t} \cdot \left(-\alpha \cdot \sin(\beta \cdot t) + \beta \cdot \cos(\beta \cdot t)\right)$$

$$i_L(t) := C \cdot V_A \cdot e^{-\alpha \cdot t} \cdot \left(-\alpha \cdot \sin(\beta \cdot t) + \beta \cdot \cos(\beta \cdot t)\right)$$

$$i_L(0) = 5.484 \times 10^{-4}$$

8-17 (D) Traditional, 15-17 Laplace Early. An inductor L is connected in parallel with a 600 Ω resistor. The parallel combination is then connected in series with a capacitor C. Select the values of L and C so that the equivalent impedance of the combination is 100 + j0 Ω at ω = 100 krad/s.

Solution: The equivalent impedance in terms of R, L and C is given by:

$$Z = \frac{j \cdot \omega \cdot L \cdot R}{j \cdot \omega \cdot L + R} + \frac{1}{j \cdot \omega \cdot C}$$

evaluation over the complex plane yields

$$Z = \omega^2 \cdot L^2 \cdot \frac{R}{\left(R^2 + \omega^2 \cdot L^2\right)} + j \cdot \left[\omega \cdot L \cdot \frac{R^2}{\left(R^2 + \omega^2 \cdot L^2\right)} - \frac{1}{(\omega \cdot C)} \right]$$

We now have the desired form to set up a solve block with two equations and two unknowns. Given and guess values must be defined.

$$R := 600 \qquad C := 10^{-9} \qquad L := 10^{-3} \qquad \omega := 10^5$$

Given

$$\omega^2 \cdot L^2 \cdot \frac{R}{\left(R^2 + \omega^2 \cdot L^2\right)} = 100$$

$$\omega \cdot L \cdot \frac{R^2}{\left(R^2 + \omega^2 \cdot L^2\right)} - \frac{1}{(\omega \cdot C)} = 0$$

$$\begin{pmatrix} L \\ C \end{pmatrix} := \text{Find}(L, C) \qquad\qquad \begin{pmatrix} L \\ C \end{pmatrix} = \begin{pmatrix} 2.683 \times 10^{-3} \\ 4.472 \times 10^{-8} \end{pmatrix}$$

Another method that may be easier and where is to define the solve block directly from the first equation for Z. If you choose to do it this way, you don't have the use the symbolic processor to get the second form.

$$C := 10^{-9} \qquad\qquad L := 10^{-3}$$

Given

$$\text{Re}\left(\frac{j \cdot \omega \cdot L \cdot R}{j \cdot \omega \cdot L + R} + \frac{1}{j \cdot \omega \cdot C} \right) = 100$$

$$\text{Im}\left(\frac{j \cdot \omega \cdot L \cdot R}{j \cdot \omega \cdot L + R} + \frac{1}{j \cdot \omega \cdot C} \right) = 0$$

$$\begin{pmatrix} L \\ C \end{pmatrix} := \text{Find}(L, C) \qquad\qquad \begin{pmatrix} L \\ C \end{pmatrix} = \begin{pmatrix} 2.683 \times 10^{-3} \\ 4.472 \times 10^{-8} \end{pmatrix}$$

The result is the same.

8-27 Traditional, 15-27 Laplace First. The circuit in Figure P8-27 is operating in the sinusoidal steady state. Use circuit reduction to find the input impedance seen by the current source and steady-state response $v_X(t)$.

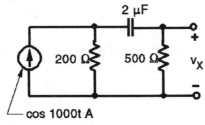

Figure P8-27

Solution: Instead of doing this problem mathematically, we will try to simulate it. This will mimic what we would have to do in the lab to verify calculations. The Orcad Capture schematic showing components and printing probes is shown in Figure P8-27sa.

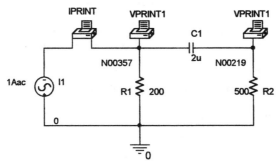

Figure P8-27sa

The Analysis setup consists of selecting the AC sweep and specifying only one frequency, $1000/2\pi = 159$ Hz. This is shown in Figure P8-27sb.

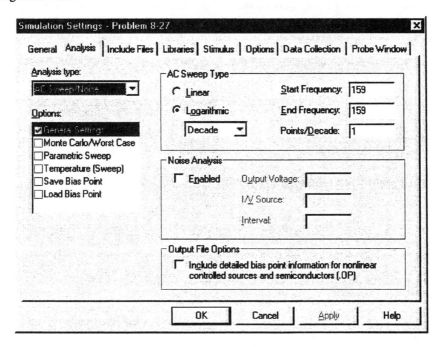

Figure P8-27sb

71

The print probes, IPRINT and VPRINT, are found under Place Parts in the Special Library. Once installed in the circuit they will report the Magnitude and Phase or/and the Real and Imaginary parts of the current (IPRINT) of the installed branch or Voltage (VPRINT) of the connected node. Before running the simulation one needs to click on the probes to bring up their Property Editor. A "y" needs to be placed in the AC, IMAG, REAL, MAG, and PHASE columns. See Figure P8-27sc.

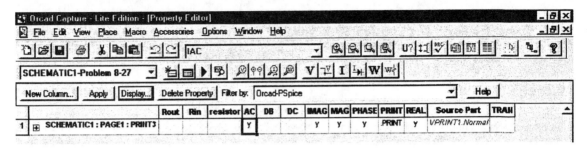

Figure P8-27sc

After running the simulation, the probe window pops up and under View select Output File. Scroll down until the results of each probe is viewed. The VPRINT and IPRINT results of this particular simulation are given in Figure P8-27sd.

FREQ   VM(N00357) VP(N00357) VR(N00357) VI(N00357)

1.590E+02   1.644E+02 -9.464E+00 **1.622E+02 -2.704E+01**

****

FREQ   IM(V_PRINT1)IP(V_PRINT1)IR(V_PRINT1)II(V_PRINT1)

1.590E+02   1.000E+00   1.590E-15   1.000E+00   2.776E-17

****

FREQ   VM(N00219) VP(N00219) VR(N00219) VI(N00219)

1.590E+02   **1.162E+02 3.556E+01** 9.453E+01   6.759E+01

Figure P8-27sd

Since the current source is $1 \angle 0°$ Ampere, the VPRINT probe at Node N00357 gives the real and imaginary parts of the input impedance directly ($Z_{IN} = V/I$): VR = 162.2 and VI = -27.1. This results in an input impedance of Z_{IN} = 162.2 - j27.1 Ω. In the meanwhile the VM and VP of the VPRINT probe at Node N00219 gives the magnitude and phase of $v_x(t)$. The values are and VM = 116.2 and VP = 35.56° giving $v_x(t) = 116.2 \cos(1000t + 35.56°)$. The IPRINT probe simply validates the value of the current source in this example.

8-33 Traditional, 15-33 Laplace Early. The circuit in Figure P8-33 is operating in the sinusoidal steady state. Use the unit output method to find the input impedance seen by the voltage source and the phasor response V_X.

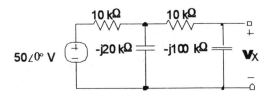

Figure P8-33

Solution: We will use the unit output method in MATLAB to find the input impedance seen by the voltage source and the phasor response V_X. The m-file is straightforward and follows:

```
function p833
%-- 4/30/03  4:26 PM --%
%First, define the components.
Z2=-j*100000;
R2=10000;
Z1=-j*20000;
R1=10000;
%Now, use the unit output method.
VX=1;
IZ2=VX/Z2;
VR2=IZ2*R2;
VZ1=1+VR2;
IZ1=VZ1/Z1;
IR1=IZ1+IZ2;
VR1=IR1*R1;
VS=VR1+VZ1;
K=1/VS;
% The input impedance can be calculated from,
ZIN=VS/IR1
Zmag=abs(ZIN)
Zarg=angle(ZIN)*180/pi
%Since we have the transfer function, K, we can calculate VX for
%a given VS.
VS=50;
VX=K*VS
VXmag=abs(VX)
VXarg=angle(VX)*180/pi
%The resulting output is:
%ZIN =1.0276e+004 -1.6690e+004i
%Zmag =1.9599e+004
%Zarg =-58.3793
%VX =34.1113 -25.1346i
%VXmag =42.3714
%VXarg =-36.3844
```

These values are compared to Orcad Capture in the next example.

8-33 Traditional, 15-33 Laplace Early. The circuit in Figure P8-33 is operating in the sinusoidal steady state. Use the unit output method to find the input impedance seen by the voltage source and the phasor response V_X.

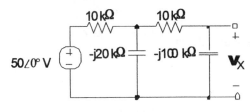

Figure P8-33

Solution: We will use Orcad Capture to simulate this circuit and compare the results with the MATLAB solution (previous page). First, we have to change the impedances to capacitances. To find these values we use the equation $C = -j/\omega Z$. We also need to specify a frequency in order to obtain a value that we can enter into Orcad. We chose 1 kHz. The capacitances are then 7.96 nF and 1.59 nF for the 20 kΩ and 100 kΩ impedances respectively. The resulting schematic appears in Figure P8-33sa.

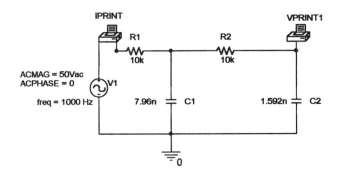

Figure P8-33sa

The magnitude of V_X is found from the VPRINT output as 42.37 V (Figure P8-33sb). The phase is given as – 36.39°. The MathLab program found. %VXmag =42.3714 %VXarg =-36.3844 – very close.

```
FREQ     VM(N00211) VP(N00211) VR(N00211)  VI(N00211)
1.000E+03  4.237E+01  -3.639E+01  3.410E+01  -2.514E+01
**********************************************************************
FREQ     IM(V_PRINT1)IP(V_PRINT1)IR(V_PRINT1)II(V_PRINT1)
1.000E+03  2.552E-03  5.837E+01  1.338E-03  2.173E-03
**********************************************************************
```
Figure P8-33sb

The input impedance magnitude and phase are derived from the source voltage and IPRINT current measurements (also Figure P8-33sb), that is, the magnitude is $|Z_{IN}| = |V_S|/|I_S| = 50/0.002552 = 19.6$ kΩ, and the phase is $\angle Z_{IN} = \angle V_S/\angle I_S = \angle 0° - \angle 58.37° = \angle -58.37°$. The current from the source leads the voltage by 58.37°. MathLab calculated ZIN as: %Zmag =1.9599e+004 %Zarg =-58.3793 – quite close.

74

8-39 Traditional, 15-39 Laplace Early (D) Design a two port circuit so that an input voltage $v_S(t) = 100$ $\cos(10^4 t)$ V delivers a steady-state output current of $i_O(t) = 10 \cos(10^4 t - 35°)$ mA to a 50-Ω resistive load.

Solution: Since the transfer function is $I_O/V_S = 0.01/100 \angle -35°$ Siemens, the network could be a simple RLR circuit shown in Figure P8-39sa.

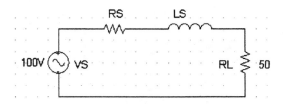

Figure P8-39sa

We can now do the design in MATLAB. The m-file follows:

```
% Define parameters of angular frequency, load, voltage source and
% output current.
w=10000;
RL=50;
VS=100;
IOmag=0.01;
IOarg=-35*pi/180;
IO=IOmag*(cos(IOarg)+j*sin(IOarg));
% Calculate the total impedance of the network.
Ztotal=VS/IO
% Calculate the impedance of the series combination of LS and RS.
Zrem=Ztotal-RL
% Calculate the value of RS, the reactance of LS, and finaly, LS.
RS=real(Zrem)
XS=imag(Zrem);
LS=XS/w
%
```

The output for the m-file follows:

Ztotal = 8.191520442889918e+003 +5.735764363510460e+003i
Zrem = 8.141520442889918e+003 +5.735764363510460e+003i
RS = 8.141520442889918e+003
LS = 0.57357643635105

8-41 Use node-voltage analysis to find the sinusoidal steady-state response $v_x(t)$ in the circuit shown in Figure P8-41.

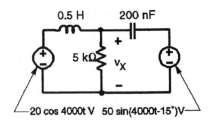

Figure P8-41

We will simulate this with Orcad Capture. The schematic is shown in Figure P8-41sa.

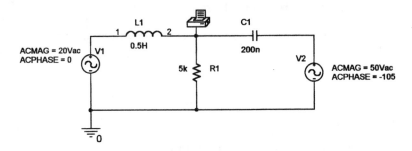

Figure P8-41sa

The Analysis setup consists of selecting the AC sweep and specifying only one frequency, $4000/2\pi = 636.6$ Hz. The print probes, IPRINT and VPRINT, are found under Place Parts in the Special Library. Once installed in the circuit they will report the Magnitude and Phase or/and the Real and Imaginary parts of the current (IPRINT) of the installed branch or Voltage (VPRINT) of the connected node. Before running the simulation one needs to click on the probes to bring up their Property Editor. A "y" needs to be placed in the AC, IMAG, REAL, MAG, and PHASE columns. See *Figure P8-27sc* for an example.

FREQ   VM(N00159) VP(N00159) VR(N00159) VI(N00159)
6.366E+02   1.211E+02   -8.409E+01   1.247E+01   -1.205E+02

Figure P8-41sb

The values from Figure P8-41sb are VM = 121.1 V and VP = -84.09° giving $v_x(t) = 121.1 \cos (4000t - 84.09°)$ V.

Figure P8-41sb

8-43 Traditional, 15-43 Laplace Early. Use node-voltage analysis to find the input impedance Z_{IN} and gain $K = V_O/V_S$ of the circuit shown in Figure P8-43 with $\mu = 100$.

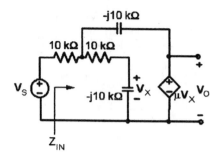

Figure P8-43

MATLAB will be used for this solution. Node A connects the two 10 kΩ resistors and node B has voltage V_X. The m-file and solution follow:

```
% Define the component resistances and reactances.
R1=10000;
R2=10000;
ZC1=-j*10000;
ZC2=-j*10000;
mu=100;
% Assume VS = 1 Volt
VS=1;
% Define the symbols
syms VA VB VO
% Write the equations using node voltage analysis.
F1=(VA-VS)/R1+(VA-VO)/ZC1+(VA-VB)/R2;
F2=(VB-VA)/R2+VB/ZC2;
F3=VO-mu*VB;
% Solve for the unknowns.
[VA,VB,VO]=solve(F1,F2,F3);
% Solve for the transfer function and input impedance.
K=VO/VS
K=double(K)
IIN=(VS-VA)/R1;
ZIN=VS/IIN
ZIN=double(ZIN)
```

The output at the command line is:

K = 100/97*i
K = 0 + 1.03092783505155i
ZIN = 19012000/1921 + 194000/1921*i
ZIN = 9.896928682977616e+003 + 1.009890681936492e+002i

8-47 Traditional, 15-47 Laplace Early. Find the input impedance Z_{IN} and gain $K = V_O/V_S$ of the circuit shown in Figure P8-47.

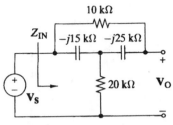

Figure P8-47

Orcad Capture is used to simulate this circuit. The capacitive reactances are converted to capacitors at a frequency of 1000 rad/sec. The circuit diagram is shown in Figure P8-47sa.

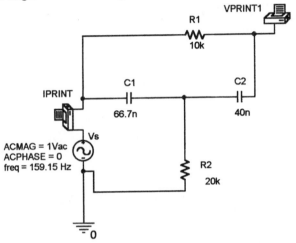

Figure P8-47sa

The current and voltage are written to the output file using the VPRINT and IPRINT probes. The print probes, IPRINT and VPRINT, are found under Place Parts in the Special Library. Once installed in the circuit they will report the Magnitude and Phase or/and the Real and Imaginary parts of the current (IPRINT) of the installed branch or Voltage (VPRINT) of the connected node. Before running the simulation one needs to click on the probes to bring up their Property Editor. A "y" needs to be placed in the AC, IMAG, REAL, MAG, and PHASE columns. See *Figure P8-27sc* for an example. The results are shown in Figure P8-47sb:

```
FREQ      IM(V_PRINT1)IP(V_PRINT1)IR(V_PRINT1)II(V_PRINT1)
1.592E+02  4.269E-05  2.447E+01  3.885E-05  1.768E-05
**********************************************************************
FREQ      VM(N00216) VP(N00216) VR(N00216) VI(N00216)
1.592E+02  8.478E-01  -1.900E+00  8.473E-01  -2.810E-02
```

Figure P8-47sb

Since we are using a 1 V source, the values for the output voltage are also the values for K, that is,
$K = 0.8478 \angle -1.9°$.

The input impedance Z_{IN} corresponds to $\mathbf{V_s}/\mathbf{I_s} = 1\angle 0°/42.69\times 10^{-6}\angle 24.47° = 23.424$ k $\angle -24.47°$ Ω or
$Z_{IN} = 21.32 \times 10^3 - j\, 9.7 \times 10^3$ Ω

8-48 Traditional, 15-48 Laplace Early. Find the sinusoidal steady state response $v_O(t)$ in the circuit shown in Figure P8-48. The element values are $V_S = 80$ mV, $R_1 = 10$ kΩ, $R_P = 5$ kΩ, $R_F = 1$ MΩ, $R_C = 10$ kΩ, $R_L = 100$ kΩ, $C = 0.25$ nF, and β = 50. The frequency is ω = 500 rad/s.

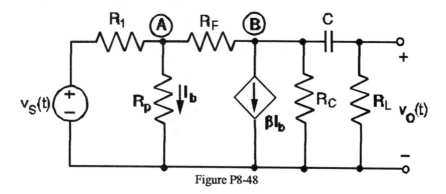

Figure P8-48

Solution: Label the node to the right of R1 A and the node to the right of R_F B as shown. We then use nodal analysis to find $v_O(t)$. The m-file and solution follow:

```
% Define the component values.
VS=0.080;
R1=10000;
Rp=5000;
RF=1000000;
RC=10000;
RL=100000;
C=0.25*10^(-9);
beta=50;
% Calculate the impedance of the capacitor.
omega=500;
ZC=1/(j*omega*C);
% Define the symbols
syms VA VB VO
% Write the equations using node voltage analysis.
FA=(VA-VS)/R1+VA/Rp+(VA-VB)/RF;
FB=(VB-VA)/RF+beta*VA/Rp+VB/RC+(VB-VO)/ZC;
FO=(VO-VB)/ZC+VO/RL;
% Solve for the unknowns.
[VA,VB,VO]=solve(FA,FB,FO);
VO=double(VO)
VOmag=abs(VO)
VOarg=angle(VO)*180/pi
```

The command line output is:

```
VO = -0.00033236501037 - 0.02474553590189i
VOmag = 0.02474776785029
VOarg = -90.76951121303196
```

We then have that $v_O(t) = 0.0247 \cos (500t - 90.77°)$.

8-61 Traditional, 15-61 Laplace Early (A) AC VOLTAGE MEASUREMENT

An ac voltmeter measurement indicates the amplitude of a sinusoid but not its phase angle. Making several measurements and using KVL can infer both the magnitude and phase. For example, Figure P8-71 shows a relay coil of unknown resistance and inductance. The following ac voltmeter reading are taken with the circuit operating in the sinusoidal steady state at $f = 1$ kHz: $|V_S| = 10$ V. $|V_1| = 4$ V, and $|V_2| = 8$ V.

(a) Use these voltage magnitude measurements to solve for R and L.

(b) Determine the phasor voltage across each element and show that the satisfy KVL.

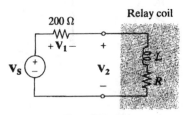

Figure P8-61

Solution: (a) Using MATLAB, we write the equations for $|V_1|$ and $|V_2|$ which will have two unknowns, R and L. We the use the solve function to determine R and L. The m-file and results follow:

```
% Define the source voltage, frequency and resistance.
VS=10;
RS=200;
f=1000;
w=2*3.14159*f;
% Define the symbols (unknowns).
syms R L
% Use the voltage divider relations for V1 and V2 to create the
% two equations and two unknowns.
F1=RS*VS/sqrt((RS+R)^2+(w*L)^2)-4;
F2=(R^2+(w*L)^2)*VS^2/((RS+R)^2+(w*L)^2)-64;
[L,R]=solve(F1,F2);
R=double(R(1))
L=double(L(1))
% Calculate V1 and V2 and verify that the satisfy KVL.
V1=VS*RS/(RS+R+j*w*L);
V2=VS*(R+j*w*L)/(RS+R+j*w*L);
V1mag=abs(V1)
V1arg=angle(V1)*180/pi
V2mag=abs(V2)
V2arg=angle(V2)*180/pi
VS-V1-V2
```

The command line output is:

```
R = 125
L = 0.06047369386827
V1mag =   4
V1arg = -49.45839812649548
V2mag =  8
V2arg = 22.33164500922151
ans =  -8.881784197001252e-016 -4.440892098500626e-016i
```

Notice that we had to look at the solutions for R and L on the command line to modify the m-file to take the first solutions for R and L. We rejected the negative solutions.

9-29 Use Mathcad or MATLAB to find the inverse transform of the following function.

$$F(s) = \frac{s \cdot \left(s^2 + 3 \cdot s + 4\right)}{(s + 2) \cdot \left(s^3 + 6 \cdot s^2 + 16 \cdot s + 16\right)}$$

Solution: This is a particularly easy problem using Mathcad. Simply copy the right hand side of the equation below, select one of the s's, select Symbolics - Transform - Inverse Laplace, and the answer appears below.
The transform:

$$\frac{s \cdot \left(s^2 + 3 \cdot s + 4\right)}{(s + 2) \cdot \left(s^3 + 6 \cdot s^2 + 16 \cdot s + 16\right)}$$

has inverse Laplace transform

$$-t \cdot \exp(-2 \cdot t) + \exp(-2 \cdot t) - \exp(-2 \cdot t) \cdot \sin(2 \cdot t)$$

We have selected "Show comments" under the Symbolics - Evaluation style to get the "has inverse Laplace transform" between the expression and its inverse transform.

9-30 Use Mathcad or MATLAB to find the inverse transform of the following function.

$$F(s) = \frac{40(s^3 + 2s^2 + s + 2)}{s(s^3 + 4s^2 + 4s + 16)}$$

Solution: We can just use the command line in MATLAB to find the inverse transform. The result follows:

» SYMS S
» ILAPLACE(40*(S^3+2*S^2+S+2)/S/(S^3+4*S^2+4*S+16))

ANS =

5+17*EXP(-4*T)+18*COS(2*T)-6*SIN(2*T)

» PRETTY(ANS)

     5 + 17 EXP(-4 T) + 18 COS(2 T) - 6 SIN(2 T)
»

9-36 The switch in Fig. P9-35 has been open for a long time. At $t = 0$ the switch is closed.
(a) Find the differential equation for the circuit and initial condition.
(b) Find $v_O(t)$ using the Laplace transformation for $v_S(t) = 10[\sin 1000t]u(t)$.
(c) Identify the forced and natural components in the response waveform and transform.

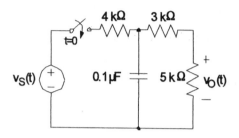

Figure P9-35

Solution: (a) For t < 0 the capacitor acts like and open circuit and $v_C(0) = 0$. For t > 0 the Thevenin circuit is $R_T = 8000 \parallel 4000 = 2666\ \Omega$ and $V_T = 0.6667\ V_S$. The circuit differential equation is then

$$R_T C \frac{d}{dt} v_C(t) + v_C(t) = V_T$$

(b) The output voltage is $v_O(t) = 5/(5+3)\ v_C(t) = 5/8\ v_C(t)$. We need to find $v_C(t)$. The Laplace transform of the equation (we have used Matlab to transform $v_S(t) = 10[\sin 1000t]u(t)$) is

$$R_T C s V_C(s) + V_C(s) = \frac{(2/3) \times 10000}{s^2 + 1000^2}$$

$$s V_C(s) + 3750 V_C(s) = 3750 \frac{6667}{s^2 + 1000^2}$$

where the MATLAB command line looks like:

%m-file
```
syms t s
laplace(0.6667*10*sin(1000*t))
```

and yields

ans =

6667/(s^2+1000000)
>>

Solve for $V_C(s)$ to get:

$$V_C(s) = \frac{37500000}{s^2 + 1000^2} \frac{1}{s + R_T C} = \frac{6667}{s^2 + 1000^2} \frac{3750}{s + 37500}$$

$$= \frac{400 \times 3750}{241 \times 1000} \frac{1000}{(s^2 + 1000^2)} - \frac{400}{241} \frac{s}{s^2 + 1000^2} + \frac{400}{241} \frac{1}{s + 3750}$$

We then must take the inverse Laplace transform to get $v_C(t)$. Finally we use a voltage divider to obtain $v_O(t)$. The MATLAB m-file looks like:

```
ilaplace((6667/(s^2+1000000))*(3750/(s+3750)))
vO=(5/8)*ilaplace((6667/(s^2+1000000))*(3750/(s+3750)))
```

The Command Screen gives back the following:

```
ans =

20001/12050*exp(-3750*t)-
20001/12050*cos(1000000^(1/2)*t)+60003/9640000*1000000^(1/2)*sin(1000000^(1/2)*
t)

vO =

20001/19280*exp(-3750*t)-
20001/19280*cos(1000000^(1/2)*t)+60003/15424000*1000000^(1/2)*sin(1000000^(1/2)
*t)
```

(c) The last two terms of $V_C(s)$ are the forced response, the first term is the natural response. For $v_O(t)$, the exponential term is the natural response and the sinusoidal terms are the forced response.

The same problem was solved using Mathcad and that yielded a much nicer format for the response:

$$L\left(\frac{1}{3750}\cdot\frac{d}{dt}v_C + v_C\right) = L(6.67\sin(1000t)\cdot u(t)) \qquad \frac{s\cdot V_C(s)}{3750} + V_C(s) = \frac{6.67\cdot1000}{s^2+1000^2}$$

Solve for $V_C(s)$ $\quad V_C(s) = \dfrac{25\cdot10^6}{\left(s^2+1000^2\right)\cdot(s+3750)} = \dfrac{400}{241\cdot(s+3750)} - \dfrac{400}{241}\cdot\dfrac{(-3750+s)}{\left(s^2+1000^2\right)}$

$$V_C(s) = \frac{400}{241\cdot(s+3750)} - \frac{400}{241}\cdot\frac{s}{s^2+1000^2} + \frac{400\cdot3750}{241\cdot1000}\cdot\frac{1000}{s^2+1000}$$

$$v_C(t) = \frac{400}{241}\cdot\exp(-3750t) + \frac{1500}{241}\cdot\sin(1000t) - \frac{400}{241}\cdot\cos(1000t) \qquad v_O(t) = v_C(t)\cdot\frac{5}{8} \quad V$$

$$v_O(t) = 1.037\exp(-3750t) + 3.89\sin(1000t) - 1.037\cos(1000t) \qquad V$$

84

9-40 The switch in Fig. P9-39 has been closed for a long time and is opened at t = 0. The circuit parameters are

$R := 5 \cdot k\Omega$ $\qquad$ $L := 10 \cdot mH$ $\qquad$ $C := 16 \cdot pF$ $\qquad$ $V_A := 10 \cdot V$

(a) Find the differential equation for the circuit and the initial conditions.
(b) Use Laplace transforms to solve for the $i_L(t)$ for $t \geq 0$.

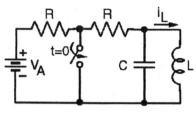

Figure P9-39

Solution: (a) Using Mathcad we proceed as follows. For t < 0 the capacitor acts like an open circuit and the inductor like a short circuit. Hence $v_C(0) = 0$, $i_L(0) = 0$, $di_L(0)/dt=1/L\, v_C(0) = 0$. For t > 0 the Norton circuit seen by the L and C in parallel is

$$i_N := \frac{V_A}{2 \cdot R} \qquad i_N = 1\,mA$$

$$R_N := 2 \cdot R \qquad R_N = 10\,k\Omega$$

The circuit differential equation is

$$L \cdot C \cdot \frac{d^2}{dt^2} i_L + \frac{L}{R_N} \frac{d}{dt} i_L + i_L = i_N$$

(b) We now take the Laplace transform of this equation to get

$$L \cdot C \cdot \left(s^2\, I_L(s) - s \cdot i_L(0) - \frac{d}{dt} i_L(0) \right) + \frac{L}{R_N} \cdot \left(s \cdot I_L(s) - i_L(0) \right) + I_L(s) = \frac{i_N}{s}$$

or

$$L \cdot C \cdot \left(s^2 I_L \right) + \frac{L}{R_N} \cdot \left(s \cdot I_L \right) + I_L = \frac{i_N}{s}$$

has solution(s)

$$\frac{i_N}{\left[s \cdot \left(L \cdot C \cdot s^2 \cdot R_N + L \cdot s + R_N \right) \right]} \cdot R_N$$

where we have used Symbolics - Variable - Solve to find $I_L(s)$. We now substitute the component values to get

$$\frac{.001}{\left[s \cdot \left(1.6 \times 10^{-13} \cdot s^2 \cdot 10^4 + 0.01 \cdot s + 10^4 \right) \right]} \cdot 10^4 \qquad\qquad L \cdot C = 1.6 \times 10^{-13}\, s^2$$

has inverse Laplace transform

$$1.00 \cdot 10^{-3} + 3.33 \cdot 10^{-4} \cdot \exp(-5000000 \cdot t) - 1.33 \cdot 10^{-3} \cdot \exp(-1250000 \cdot t)$$

where we have used Symbolics - Transform - Inverse Laplace and have truncated the coefficients of the result to 3 significant figures.

9-51 (A) THE DOMINANT POLE APPROXIMATION

When a transform $F(s)$ has widely separated poles, then those closest to the j-axis tend to dominate the response because they have less damping. An approximation to the waveform can be obtained by ignoring the contributions of all except the dominant poles. We can ignore the nondominant poles by simply discarding their terms in the

$$F(s) = 10^6 \frac{s+1000}{(s+4000)[(s+25)^2 + 100^2]}$$

partial fraction expansion of $F(s)$. The purpose of this example is to examine a dominant pole approximation. Consider the transform

(a) Construct a partial-fraction expansion of $F(s)$ and find $f(t)$.
(b) Construct a pole-zero diagram of $F(s)$ and identify the dominant poles.
(c) Construct a dominant pole approximation $g(t)$ by discarding the nondominant poles in the partial fraction expansion in (a).
(d) Plot $f(t)$ and $g(t)$ and comment on the accuracy of the approximation.

Solution: (a) The partial-fraction expansion of $F(s)$ is easily accomplished and is given by

$$F(s) = \frac{-189.7}{s+4000} + \frac{94.87 - 1228j}{s+25-100j} + \frac{94.87+1228j}{s+25+100j}$$

where we used the m-file that follows:

```
% Find the coefficients of the partial fraction expansion using
% equation (9-20) of your text.
K=10^6;
z1=-1000;
p1=-4000;
p2=-25+j*100;
p3=-25-j*100;
k1=K*(p1-z1)/(p1-p2)/(p1-p3);
k2=K*(p2-z1)/(p2-p1)/(p2-p3);
k3=K*(p3-z1)/(p3-p2)/(p3-p1);
%
```

which gave the following results:

```
k1 = -1.897458196624106e+002
k2 =  9.487290983120529e+001 -1.228801834209590e+003i
k3 =  9.487290983120528e+001 +1.228801834209590e+003i
```

so that

$$f(t) = -189.7e^{-4000t} + 2Ae^{-25t}\cos(100t+\phi)$$
$$A = 2|94.87 - 1228j| \quad and \quad \phi = \arg(94.87 - 1228j)$$

(b) The poles and zeros for the transform are easily shown to be $z_1 = -1000$, $p_1 = -4000$, $p_c = -25$ G j 100. They can be easily plotted in MATLAB once the real and imaginary parts of each have been defined. The command to plot them is then

```
plot(RZ1,IZ1,'o',RP1,IP1,'x',RPC1,IPC1,'x',RPC2,IPC2,'o')
```

The resulting plot is shown in Figure P9-51sa. The dominant poles are the complex conjugate pair.

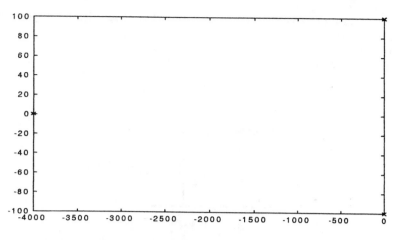

Figure P9-51sa

(c) After we discard the non-dominant pole of (a) we have

$$G(s) = \frac{94.87 - 1228j}{s + 25 - 100j} + \frac{94.87 + 1228j}{s + 25 + 100j}$$

For this approximation, we have

$$g(t) = 2Ae^{-25t} \cos(100t + \phi)$$

The remainder of the m-file follows along with the graph of the two functions.

```
% Now plot the functions f(t) and g(t)
t=0:.02:.25;
A=2*abs(k2);
phi=angle(k2);
f=k1*exp(p1*t)+2*A*exp(real(p2)*t).*cos(imag(p2)*t+phi);
g=2*A*exp(real(p2)*t).*cos(imag(p2)*t+phi);
plot(t,f,'-',t,g,'--')
%
```

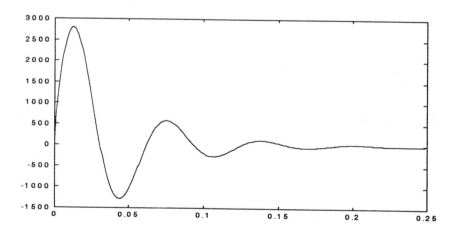

10-39 (D) The circuit in Figure P10-39 represents an active RC highpass filter originally proposed by R. P. Sallen and E. K. Key in 1955.
(a) Transform the circuit into the s domain and find the circuit determinant.
(b) Select values of R, C, and μ so the circuit has natural poles at $s = \pm j5000$ rad/s.

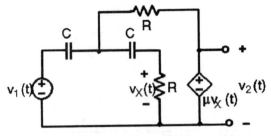

Figure P10-39

Solution: (a) Let the node between the capacitors be node A and the node at $v_X(t)$ be node B. The equation at node A is

$$\left(V_A - V_1\right)\cdot C\cdot s + \left(V_A - V_B\right)\cdot C\cdot s + \left(V_A - \mu\cdot V_B\right)\cdot\frac{1}{R} = 0$$

The equation at node B is

$$\frac{1}{R}\cdot V_B + \left(V_B - V_A\right)\cdot C\cdot s = 0$$

Rearranging these two equations gives

$$\left(\frac{1}{R} + 2C\cdot s\right)\cdot V_A - \left(\frac{\mu}{R} + C\cdot s\right)\cdot V_B = C\cdot s\cdot V_1$$

$$-C\cdot s\cdot V_A + \left(\frac{1}{R} + C\cdot s\right)\cdot V_B = 0$$

or

$$\begin{bmatrix} \frac{1}{R} + 2C\cdot s & -\left(\frac{\mu}{R} + C\cdot s\right) \\ -C\cdot s & \frac{1}{R} + C\cdot s \end{bmatrix}\cdot\begin{pmatrix} V_A \\ V_B \end{pmatrix} = \begin{pmatrix} C\cdot s \\ 0 \end{pmatrix}$$

The matrix determinant can be found by selecting the matrix and selecting Symbolics - Matrix - Determinant

$$\begin{bmatrix} \frac{1}{R} + 2C\cdot s & -\left(\frac{\mu}{R} + C\cdot s\right) \\ -C\cdot s & \frac{1}{R} + C\cdot s \end{bmatrix}$$

has determinant

$$\frac{1 + 3\cdot C\cdot s\cdot R + C^2\cdot s^2\cdot R^2 - C\cdot s\cdot R\cdot\mu}{R^2} = \frac{(R\cdot C)^2\cdot s^2 + (3 - \mu)\cdot R\cdot C\cdot s + 1}{R^2}$$

Poles at $\pm j$ 5000 axis require that

$$\mu := 3 \qquad \text{and} \qquad RC := \frac{1}{5000}\cdot\text{sec}$$

If we choose $C := 20\cdot nF$

Then R is found by $\qquad R := \frac{RC}{C} \qquad R = 10\,k\Omega$

88

10-54 (A D) LIGHTING PULSER DESIGN

The circuit in Figure P10-54sa is a simplified circuit diagram of a pulser that is used to test equipment against lightning-induced transients. Select the values of V_0, R_1, R_2, C_1, and C_2 so that the pulser delivers an output pulse

$$v_O(t) = 500(e^{-2000t} - e^{-80000t}) \text{ V}$$

The pulser is completely discharged prior to delivering a pulse.

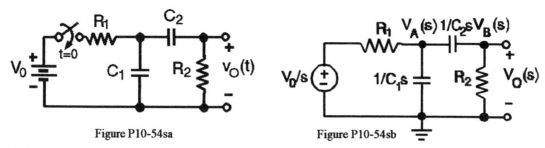

Figure P10-54sa   Figure P10-54sb

Solution: First, transform $v_O(t)$ into the s domain (Symbolics - Transform - Laplace) - see Figure P10-54sb. Node A is taken at the junction of R_1 and C_1. Node B is at the junction of R_2 and C_2.

$$500 \cdot \left(e^{-2000 \cdot t} - e^{-80000 \cdot t} \right)$$

has Laplace transform

$$\frac{500}{(s + 2000)} - \frac{500}{(s + 80000)}$$

by factoring, yields

$$\frac{39000000}{[(s + 2000) \cdot (s + 80000)]}$$

expands to

$$\frac{\dfrac{39000000}{160000000}}{\left(\dfrac{s^2 + 82000 \cdot s + 160000000}{160000000} \right)}$$

yields

$$\frac{\dfrac{39}{160}}{\dfrac{1}{160000000} \cdot s^2 + \dfrac{41}{80000} \cdot s + 1}$$

$$\frac{0.244}{6.25 \cdot 10^{-9} \cdot s^2 + 5.125 \cdot 10^{-4} \cdot s + 1}$$

A number of symbolic manipulations are done. Some editing steps are not shown.

89

Next we solve the circuit for V_B

Writing two node voltage equations

$$\begin{pmatrix} C_1 \cdot s + C_2 \cdot s + \dfrac{1}{R_1} & -C_2 \cdot s \\ \\ -C_2 \cdot s & C_2 \cdot s + \dfrac{1}{R_2} \end{pmatrix} \cdot \begin{pmatrix} V_A \\ V_B \end{pmatrix} = \begin{pmatrix} \dfrac{V_0}{R_1 \cdot s} \\ \\ 0 \end{pmatrix}$$

Solving for $V_O(s) = V_B$

$$\Delta(s) = \frac{\left(C_1 \cdot s^2 \cdot R_1 \cdot C_2 \cdot R_2 + C_1 \cdot s \cdot R_1 + C_2 \cdot s \cdot R_1 + C_2 \cdot s \cdot R_2 + 1 \right)}{\left(R_1 \cdot R_2 \right)} \quad ; \quad \Delta_B(s) = \frac{C_2 \cdot V_0}{R_1}$$

$$V_O(s) = \frac{\Delta_B(s)}{\Delta(s)} = \frac{R_2 \cdot C_2 \cdot V_0}{R_1 \cdot R_2 \cdot C_1 \cdot C_2 \cdot s^2 + \left(R_1 \cdot C_1 + R_2 \cdot C_2 + R_1 \cdot C_2 \right) \cdot s + 1}$$

Comparing this result to the required transform

$$V_O(s) = \frac{0.244}{6.25 \cdot 10^{-9} \cdot s^2 + 5.125 \cdot 10^{-4} \cdot s + 1} \qquad \text{yields the following design constraints:}$$

$V_0 := 1000$

$TC_2 := \dfrac{39}{160 \cdot V_0} \qquad TC_2 = 2.437 \times 10^{-4}$

$TC_1 := \dfrac{1}{160000000 \cdot TC_2} \qquad TC_1 = 2.564 \times 10^{-5}$

$TC_{12} := -\left(TC_1 + TC_2 \right) + \dfrac{41}{80000} \qquad TC_{12} = 2.431 \times 10^{-4}$

Let $\quad R_1 := 10000 \quad$ Then we find C_1, C_2 and R_2 completing our design.

$C_1 := \dfrac{TC_1}{R_1} \qquad\qquad C_2 := \dfrac{TC_{12}}{R_1} \qquad\qquad R_2 := \dfrac{TC_2}{C_2}$

$C_1 = 2.564 \times 10^{-9} \qquad C_2 = 2.431 \times 10^{-8} \qquad R_2 = 1.003 \times 10^4$

11-9 Transform the circuit in Figure P11-9 into the s domain and solve for the transfer function $T_V(s) = V_2(s)/V_1(s)$. Locate the poles and zeros of the transfer function.

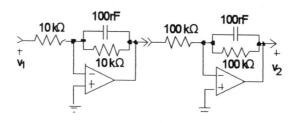

Figure P11-9

Solution: We will use Mathcad to solve this circuit problem.

$$Z_2(s) = \frac{1}{10^{-7} \cdot s + \frac{1}{10^4}} = \frac{10^7}{(s + 1000)}$$

$$Z_2(s) = \frac{1}{10^{-7} \cdot s + \frac{1}{10^5}} = \frac{10^7}{(s + 100)}$$

$$Z_1(s) = 10^4 \quad T_{V1}(s) = \frac{Z_2}{Z_1} = \frac{-1000}{s + 1000}$$

$$Z_1(s) = 10^5 \quad T_{V2}(s) = \frac{Z_2}{Z_1} = \frac{-100}{s + 100}$$

Using the chain rule

$$T_V(s) = T_{V1}(s) \cdot T_{V2}(s) = \left(\frac{-1000}{s + 1000}\right) \cdot \left(\frac{-100}{s + 100}\right)$$

$T_V(s)$ has poles at s = -1000, s = -100 and a double zero at infinity

11-49 The step response of a linear circuit is $g(t) = 10[e^{-1000t} \sin 2000t] \, u(t)$. Find the sinusoidal steady-state response for the input $x(t) = 10 \cos 2000t$.

Solution: Our approach here is to take Laplace transform of the step response and multiply by s to get the transfer function. We then evaluate the transfer function at $\omega = 2000$ to get the transfer function at that frequency. Then multiply the input to get the response.

Take the Laplace transform of the step response.

$$10 \cdot e^{-1000 \cdot t} \cdot \sin(2000 \cdot t)$$

has Laplace transform

$$\frac{20000}{\left[(s + 1000)^2 + 4000000\right]}$$

Multiply this by s and substitute $j\omega$ for s to get the transfer function.

$$t(\omega) := \frac{20000 \cdot j \cdot \omega}{\left[(j \cdot \omega + 1000)^2 + 4000000\right]}$$

Evaluate this at $\omega = 2000$.

$$|t(2000)| = 9.701$$

$$\arg(t(2000)) = 14.036 \, deg$$

The magnitude and phase of the response for $x(t) = 10 \cos 2000t$ is then

$$A := 10 \cdot |t(2000)| \qquad \phi := \arg(t(2000))$$

$$A = 97.014 \qquad \phi = 14.036 \, deg$$

So the sinusoidal steady-state response is

$$97.014 \cdot \cos[2000(t - 14.036) \cdot deg]$$

11-51 (D) Design a circuit to realize the transfer function below using only resistors, capacitors, and OP AMPs. Scale the circuit so that all resistors are greater than 10 kΩ and all capacitors are less the 1 μF.

$$T_V(s) = \pm \frac{10^5}{(s+200)(s+2500)}$$

Solution: Partition the transfer function and re-write as

$$T_V(s) = \frac{200/s}{1+200/s} \frac{500/s}{1+2500/s} = \frac{Z_2(s)}{Z_1(s)+Z_2(s)} \frac{Z_4(s)}{Z_3(s)+Z_4(s)}$$

Now we have the product of two voltage dividers with impedances $Z_1(s) = 1$, $Z_2(s) = 200/s$, $Z_3(s) = 1 + 2000/s$ and $Z_4(s) = 500/s$. In terms of prototype components, we have $Z_1(s)$ is a resistance $R_1 = 1$, $Z_2(s)$ is a capacitance $C_2 = 1/200$, $Z_3(s)$ is a resistance in series with a capacitance, $R_3 = 1$ and $C_3 = 1/2000$, and $Z_4(s)$ is a capacitance $C_4 = 1/500$.

We now need to scale the components so that all resistors are greater than 10 kΩ and all capacitors are less the 1 μF. A suitable scale factor would be 10^5. We then have for our components $R_1 = 100$ kΩ, $C_2 = 50$ nF, $R_3 = 100$ kΩ, $C_3 = 5$ nF, and $C_4 = 20$nF. Of course, we need to isolate the two stages with an OP AMP follower circuit.

We now go to Electronics Workbench to simulate the circuit. The schematic is shown in Figure P11-51sa. The sinusoidal steady state frequency response is shown in Figure P11-51sb.

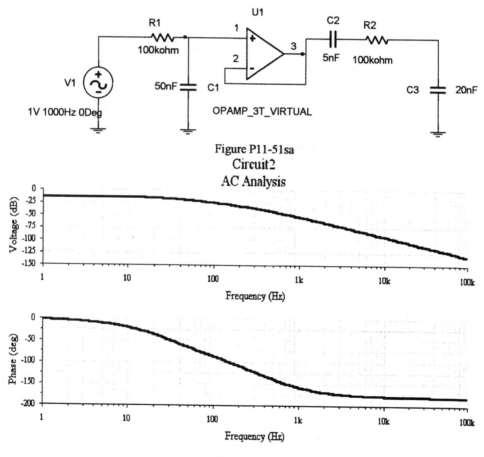

Figure P11-51sa

Circuit2

AC Analysis

Figure P11-51sb

93

11-52 (D) Design a circuit to realize the transfer function below using only resistors, capacitors and not more than one OP AMP. Scale the circuit so that all capacitors are exactly 100 pF.

$$T_V(s) = \pm \frac{100\,(s+500)}{(s+200)(s+2500)}$$

Solution: Partition the function as

$$T_V(s) = \frac{100}{s+200}\frac{s+500}{s+2500} = \frac{Y_1(s)}{Y_1(s)+Y_2(s)}\frac{Y_3(s)}{Y_3(s)+Y_4(s)}$$

Now we have the product of two voltage dividers with admittances $Y_1(s) = 100$, $Y_2(s) = s + 100$, $Y_3(s) = s + 500$ and $Y_4(s) = 2000$. In terms of prototype components, we have $Y_1(s)$ is a resistance $R_1 = 1/100$, $Y_2(s)$ is a capacitor $C_2 = 1$ in parallel with a resistor $R_2 = 1/100$, $Y_3(s)$ is a capacitor $C_3 = 1$ in parallel with a resistor $R_3 = 1/500$, and $Y_4(s)$ is a resistance $R_4 = 1/2000$.

We now need to scale the components so that all capacitors are exactly 100 pF. The scale factor must be 10^{10}. We then have for our resistors $R_1 = 100$ MΩ, $R_2 = 100$ MΩ, $R_3 = 20$ MΩ, $R_4 = 5$ MΩ. Of course, we need to isolate the two stages with an OP AMP follower circuit.

We now go to Orcad Capture to simulate the circuit. The schematic is shown in Figure P11-52sa.

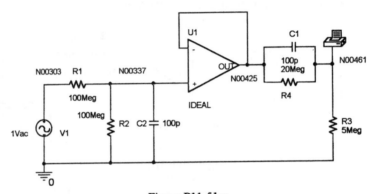

Figure P11-51sa

The VPRINT1 device is used to capture the magnitude and phase of the output voltage (Node N00462 in this run). The pertinent portion of the Output File is shown below.

FREQ	VP(N00461)	VDB(N00461)
1.000E+00	-1.223E+00	-2.000E+01
1.585E+00	-1.938E+00	-2.001E+01
2.512E+00	-3.066E+00	-2.002E+01
3.981E+00	-4.838E+00	-2.006E+01
6.310E+00	-7.587E+00	-2.014E+01
1.000E+01	-1.172E+01	-2.034E+01
1.585E+01	-1.749E+01	-2.080E+01
2.512E+01	-2.437E+01	-2.171E+01
3.981E+01	-3.049E+01	-2.316E+01
6.310E+01	-3.383E+01	-2.492E+01
1.000E+02	-3.496E+01	-2.651E+01
1.585E+02	-3.702E+01	-2.779E+01
2.512E+02	-4.262E+01	-2.907E+01
3.981E+02	-5.175E+01	-3.083E+01
6.310E+02	-6.206E+01	-3.336E+01
1.000E+03	-7.103E+01	-3.658E+01
1.585E+03	-7.763E+01	-4.022E+01
2.512E+03	-8.209E+01	-4.407E+01

3.981E+03	-8.498E+01	-4.801E+01
6.310E+03	-8.683E+01	-5.198E+01
1.000E+04	-8.800E+01	-5.597E+01
1.585E+04	-8.873E+01	-5.997E+01
2.512E+04	-8.920E+01	-6.396E+01
3.981E+04	-8.950E+01	-6.796E+01
6.310E+04	-8.968E+01	-7.196E+01
1.000E+05	-8.980E+01	-7.596E+01
1.585E+05	-8.987E+01	-7.996E+01
2.512E+05	-8.992E+01	-8.396E+01
3.981E+05	-8.995E+01	-8.796E+01
6.310E+05	-8.997E+01	-9.196E+01
1.000E+06	-8.998E+01	-9.596E+01

The sinusoidal steady state frequency response of the Magnitude in dB and the Phase in degrees is shown in Figure P11-52sb.

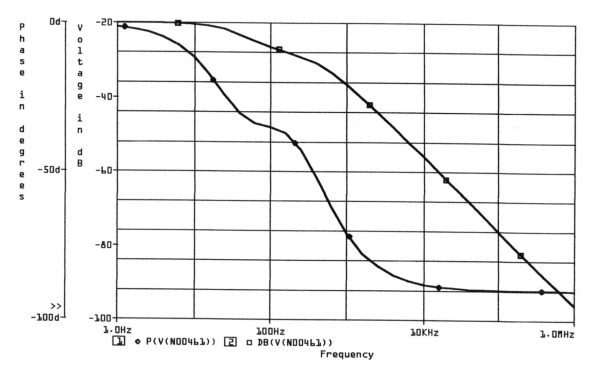

Figure P11-51sb

95

11-55 (D) Design a circuit to realize the following transfer function using only resistors, capacitors, and no more than one OP AMP. Scale the circuit so that all resistors are greater than 10 kΩ and all capacitors are less than 1 μF.

$$T_V(s) = \frac{(s+100)(s+1000)}{(s+200)(s+500)}$$

Solution: Partition the transfer function as

$$T_V(s) = \left[\frac{(s+100)}{(s+200)}\right][2]\left[\frac{(s+1000)}{(2s+1000)}\right]$$

The first stage voltage divider has

$$T_{V1}(s) = \frac{(s+100)}{(s+200)} = \frac{Y_1}{Y_1+Y_2}$$

so that $Y_1 = s + 100$ (a capacitor in parallel with a resistor) and $Y_2 = 100$ (a resistor). The scaling factor for this stage will be 10^7, which gives resistor values of 100 kΩ and a capacitor value of 0.1 μF. The second stage is an OP AMP stage with a gain of 2. For maximum input impedance (no loading to stage 1) we use a non-inverting amplifier. The third stage voltage divider has

$$T_{V2}(s) = \frac{(s+1000)}{(2s+1000)} = \frac{Y_3}{Y_3+Y_4}$$

so that $Y_3 = s + 1000$ (a capacitor in parallel with a resistor) and $Y_4 = s$ (a capacitor). Again, the scaling factor for this stage will be 10^7, which gives a resistor value of 10 kΩ and capacitor values of 0.1 μF. The resulting schematic (realized in Electronic Workbench) is shown in Figure P11-55sa. The simulation results are shown in Figure P11-55sb.

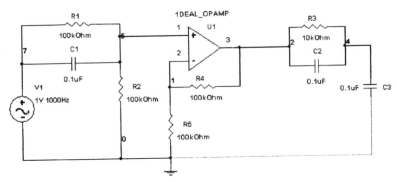

Figure P11-55sa

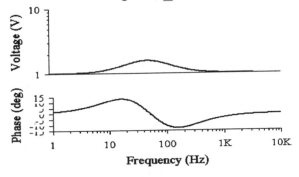

Figure P11-55sb

11-62 (A D) SECOND-ORDER CIRCUIT STEP RESPONSE

The step response in Figure P11-62 is of the form

$$g(\alpha,\beta,K,t) := \frac{K}{\alpha^2 + \beta^2} \cdot \left(1 - e^{-\alpha \cdot t} \cdot \cos(\beta \cdot t) - \frac{\alpha}{\beta} e^{-\alpha \cdot t} \cdot \sin(\beta \cdot t) \right)$$

(a) Estimate values for the parameters K, α, and β from the plot of $g(t)$.
(b) Find the transfer function $T(s)$ corresponding to $g(t)$.
(c) Design a circuit that realizes the $T(s)$ found in (b).
(d) Use computer-aided circuit analysis to verify your design.

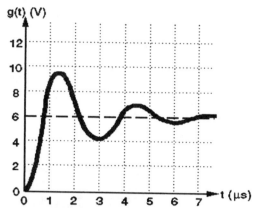

Solution: (a) We have three unknows so we need to generate threee equations. We determine that $K/(\alpha^2 + \beta^2) = 6$ since the function goes to 6 as t goes to infinity. For the other two equations, we choose two points on the graph and put the resulting equations into the solve block. Here it is, along with some educated guesses.

$$\beta := \frac{2 \cdot \pi}{3 \cdot 10^{-6}} \qquad \alpha := \frac{1}{4 \cdot 10^{-6}}$$

Given

$$6 \cdot \left(1 - e^{-\alpha \cdot 10^{-6}} \cdot \cos(\beta \cdot 10^{-6}) - \frac{\alpha}{\beta} e^{-\alpha \cdot 10^{-6}} \cdot \sin(\beta \cdot 10^{-6}) \right) = 8.25$$

$$6 \cdot \left(1 - e^{-\alpha \cdot 3 \cdot 10^{-6}} \cdot \cos(\beta \cdot 3 \cdot 10^{-6}) - \frac{\alpha}{\beta} e^{-\alpha \cdot 3 \cdot 10^{-6}} \cdot \sin(\beta \cdot 3 \cdot 10^{-6}) \right) = 4$$

$$\begin{pmatrix} \alpha \\ \beta \end{pmatrix} := \text{Find}(\alpha, \beta) \qquad \begin{pmatrix} \alpha \\ \beta \end{pmatrix} = \begin{pmatrix} 3.449 \times 10^5 \\ 2.272 \times 10^6 \end{pmatrix}$$

$$K := 6 \cdot (\alpha^2 + \beta^2) \qquad K = 3.169 \times 10^{13}$$

As a check, let's plot the result.

$$t := 0, 10^{-8} .. 7 \cdot 10^{-6}$$

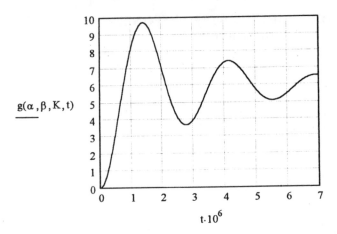

$$g(\alpha, \beta, K, t)$$

(b) We now user the Symbolic processor to take the Laplace transform of $g(t)$ and simplify the result.

$$\frac{K}{\alpha^2 + \beta^2} \cdot \left(1 - e^{-\alpha \cdot t} \cdot \cos(\beta \cdot t) - \frac{\alpha}{\beta} e^{-\alpha \cdot t} \cdot \sin(\beta \cdot t)\right)$$

has Laplace transform

$$\frac{-K}{\left[(\alpha^2 + \beta^2) \cdot \beta\right]} \cdot \left[\frac{-\beta}{s} + \beta \cdot \frac{(s + \alpha)}{\left[(s + \alpha)^2 + \beta^2\right]} + \alpha \cdot \frac{\beta}{\left[(s + \alpha)^2 + \beta^2\right]}\right]$$

simplifies to

$$\frac{K}{\left[\left(s^2 + 2 \cdot s \cdot \alpha + \alpha^2 + \beta^2\right) \cdot s\right]}$$

Since this is the step response, we have that the transfer function is

$$G(s) = \frac{K}{s^2 + 2 \cdot s \cdot \alpha + \alpha^2 + \beta^2}$$

(c) Partition $G(s)$ into two stages.

$$G(s) = \frac{K}{s^2 + 2 \cdot s \cdot \alpha + \alpha^2 + \beta^2} = \frac{K}{\alpha^2 + \beta^2} \cdot \frac{\alpha^2 + \beta^2}{s^2 + 2 \cdot s \cdot \alpha + \alpha^2 + \beta^2} = \frac{K}{\alpha^2 + \beta^2} \cdot \frac{\dfrac{\alpha^2 + \beta^2}{s}}{s + 2 \cdot \alpha + \dfrac{\alpha^2 + \beta^2}{s}}$$

The first stage is a non-inverting amplifier with gain $K/(\alpha^2 + \beta^2) = 6$. The second stage can be an RLC voltage divider with

$$\frac{Z_2}{Z_1 + Z_2} = \frac{\dfrac{\alpha^2 + \beta^2}{s}}{s + 2 \cdot \alpha + \dfrac{\alpha^2 + \beta^2}{s}}$$

where Z2 is a capacitor, and Z1 is an inductor in series with a resistor. The prototype values of these components are

$$C_p := \frac{1}{\alpha^2 + \beta^2} \cdot F \qquad C_p = 1.893 \times 10^{-13} F$$

$$R_p := 2 \cdot \alpha \cdot \Omega \qquad\qquad R_p = 6.898 \times 10^5 \Omega$$

$$L_p := 1 \cdot H \qquad\qquad L_p = 1\,H$$

The 1 H inductor is a little large. A suitable scale factor might be 10^{-3}. After application of the scale factor the components are

$$k_m := 10^{-3}$$

$$C := \frac{C_p}{k_m} \qquad C = 189.306\,pF \qquad L := k_m \cdot L_p \qquad L = 1\,mH \qquad R := k_m \cdot R_p \qquad R = 689.815\,\Omega$$

The circuit is simulated in Orcad Capture. Figures P11-62sa and P11-62sb show the schematic and the simulation results

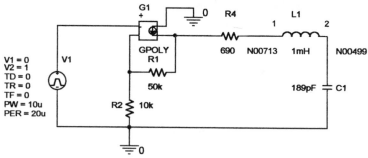

Figure P11-62sa

Figure P11-62sb

99

12-13 (a) Find the transfer function $T_V(s) = V_2(s)/V_1(s)$ for the *RLC* circuit in Figure P12-13

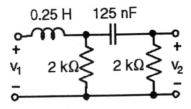

Figure P12-13

(b) Determine the dc gain, infinite frequency gain, damping ratio ζ, and undamped natural frequency ω_0. Is the gain response low pass, high pass, or bandpass?

(c) Draw the straight-line approximation of the gain of $T_V(j\omega)$.

(d) Use the straight-line approximation to calculate the gain at $\omega = 0.5\omega_0$, ω_0, and $2\omega_0$. Compare these straight-line gains with the actual values of $|T_V(j\omega)|$ at the same frequencies.

Solution: (a) Let the node at the junction of *R*, *L* and *C* be *V* and let $V_1 = 1$. Nodal analysis yields the following matrix equation:

$$\begin{bmatrix} \dfrac{1}{sL} + \dfrac{1}{R} + sC & -sC \\[2mm] -sC & \dfrac{1}{R} + sC \end{bmatrix} \begin{bmatrix} V \\[2mm] V_2 \end{bmatrix} = \begin{bmatrix} \dfrac{1}{sL} \\[2mm] 0 \end{bmatrix}$$

The solution[1] for V_2 (which is $T_V(s)$) after substituting for *R*, *L* and *C* is:

$$T_V(s) = \frac{4000\,s}{s^2 + 6000\,s + 16000000} = \frac{K\,s}{s^2 + 2\zeta\,\omega_0\,s + \omega_0^2}$$

(b) This is a second order frequency response. We can compare the expressions in the equation above to arrive at

$\omega_0 = 4000$

$2\zeta\,\omega_0 = 6000$ yields $\zeta = 3/4$

dc gain is 0

∞ gain is 0

The gain response is bandpass.

(c) Since we want to use MATLAB for this problem, we will use the bode function to make this plot. We only need to define a vector of coefficients for the numerator and one for the denominator. The command line display follows:

```
» num=[0,4000,0]
num =
        0        4000        0
```

---

[1] Using MATLAB for this symbolic solution yielded
118059162071741150000*s/(295147905179352875*s^2+1770887431076117151712*s+47223664828696452136
96000) for this transfer function. Analysis using the symbolic processor in Mathcad yielded the nicer solution presented here.

```
» den=[1 6000 16000000]
den =
            1          6000       16000000
» bode(num,den)
```

The last line generates the desired graph.

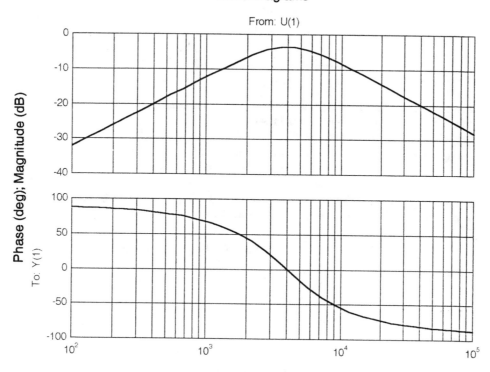

Figure P12-13sa

(d) The approximations from the figure are for $\omega = 0.5\omega_0$, ω_0, and $2\omega_0$, $T_V(s) = $ -6 dB, -3 dB (the straight-line approximation would be 0 dB), and -6 dB respectively. MATLAB yields $T_V(s) = $ -7.52 dB, -4.05 dB, and -7.52 dB for these values.

12-15 (a) Find the transfer function $T_V(s) = V_2(s)/V_1(s)$ for the OP AMP circuit in Figure P12-15.

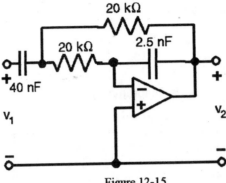

Figure 12-15

(b) Determine the dc gain, infinite frequency gain, damping ratio ζ, and undamped natural frequency ω_0. Is the gain response low pass, high pass, or bandpass?

(c) Draw the straight-line approximation of the gain of $T_V(j\omega)$.

(d) Use the straight-line approximation to calculate the gain at $\omega = 0.5\omega_0$, ω_0, and $2\omega_0$. Compare these straight-line gains with the actual values of $|T_V(j\omega)|$ at the same frequencies.

Solution: Using nodal analysis we (node A is at the junction of the 40 nF capacitor and the 20 kΩ resistors and node B is at the inverting terminal of the OP AMP) can set up a symbolic solve block. We also set $V_1 = 1$.

$$R := 20000 \qquad C_1 := 40 \cdot 10^{-9} \qquad C_2 := 2.5 \cdot 10^{-9} \qquad V_B := 0$$

Given

$$\left(V_A - 1\right) \cdot s \cdot C_1 + \frac{V_A}{R} + \frac{V_A - V_2}{R} = 0$$

$$\frac{V_B - V_A}{R} + \left(V_B - V_2\right) \cdot s \cdot C_2 = 0$$

$$\text{Find}\left(V_2, V_A\right) \rightarrow \begin{bmatrix} -20000 \cdot \dfrac{s}{\left(25000000. + s^2 + 2500 \cdot s\right)} \\[4mm] \dfrac{s^2}{\left(25000000. + s^2 + 2500 \cdot s\right)} \end{bmatrix}$$

So we have

$$T_V(s) := \frac{-20000s}{s^2 + 2500 \cdot s + 25000000}$$

(b) For the dc gain and infinite frequency gain we have

$$\left|T_V(0)\right| = 0 \qquad\qquad \left|T_V\left(10^{100}\right)\right| = 0$$

By inspection, $\quad \omega_0 := \sqrt{25000000} \quad \omega_0 = 5 \times 10^3 \qquad f_0 := \dfrac{\omega_0}{2 \cdot \pi} \qquad \zeta := \dfrac{2500}{2 \cdot \omega_0} \qquad \zeta = 0.25$

The response is bandpass. $\qquad f_0 = 795.775$

102

(c) For this part of the exercise, we want to use PSpice rather than do the approximations. The circuit and results follow:

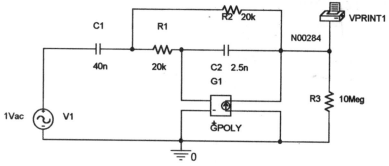

Figure P12-15sa

Notice that a 10 MΩ resistor or larger is needed at the output. This resistor is sufficiently large to appear as an open circuit since it does not load the circuit being analyzed. However, if this resistor were not there, an error in the analysis occurs stating that the output node of the dependent current source is floating and the simulation would not proceed. The dB voltage probe gives the following graph. The cursor was used to find the peak value and the center frequency, f_0.

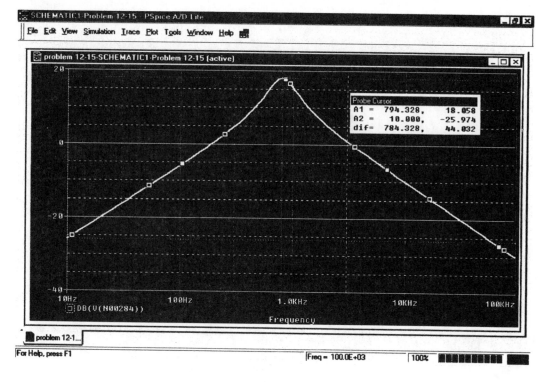

Figure P12-15sb

103

(d) The exact values of the magnitude of the transfer function at $\omega = 0.5\omega_0$, ω_0, and $2\omega_0$ are

$$20 \cdot \log\left(\left|T_V\left(0.5 \cdot j \cdot \omega_0\right)\right|\right) = 8.062$$

$$20 \cdot \log\left(\left|T_V\left(j\omega_0\right)\right|\right) = 18.062$$

$$20 \cdot \log\left(\left|T_V\left(2j\omega_0\right)\right|\right) = 8.062$$

12-16 (a) Find the transfer function $T_V(s) = V_2(s)/V_1(s)$ for the OP AMP circuit in Figure P12-16 for $R_1 = 10$ kΩ, $R_2 = 40$ kΩ, and $C = 20$ nF.

(b) Determine the dc gain, infinite-frequency gain, the damping ratio ζ, and the undamped natural fr ω_0. Is the gain response low-pass, high-pass, or bandpass?

(c) Draw the straight-line approximation of the gain of $T_V(j\omega)$.

(d) Use the straight-line gain approximation to estimate the amplitude of the steady state output for a 0.5-V sinusoidal input at $\omega = 0.5\omega_C$, ω_C, and $2\omega_C$.

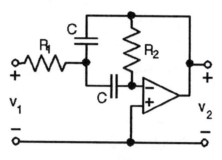

Figure P12-16

Solution: Using nodal analysis we (node A is at the junction of the capacitors and R_1 and node B is at the inverting terminal of the OP AMP) can set up a symbolic solve block. We also set $V_1 = 1$.

$$R_1 := 10000 \qquad R_2 := 40000 \qquad C := 20 \cdot 10^{-9} \qquad V_B := 0$$

Given

$$\frac{(V_A - 1)}{R_1} + (V_A - V_2)\cdot s \cdot C + (V_A - V_B)\cdot s \cdot C = 0$$

$$(V_B - V_A)\cdot s \cdot C + \frac{(V_B - V_2)}{R_2} = 0$$

$$\text{Find}(V_2, V_A) \rightarrow \begin{bmatrix} \dfrac{-5000}{\left(6250000 + 2500\cdot s + s^2\right)} \cdot s \\ \dfrac{6250000}{\left(6250000 + 2500\cdot s + s^2\right)} \end{bmatrix}$$

So we have

$$T_V(s) := \frac{-5000\cdot s}{\left(6250000 + 2500\cdot s + s^2\right)}$$

(b) For the dc gain and infinite frequency gain we have

$$|T_V(0)| = 0 \qquad\qquad \left|T_V(10^{100})\right| = 0$$

By inspection, $\quad \omega_0 := \sqrt{6250000} \qquad \omega_0 = 2.5 \times 10^3 \qquad f_0 := \dfrac{\omega_0}{2\cdot\pi} \qquad \zeta := \dfrac{2500}{2\cdot\omega_0} \qquad \zeta = 0.5$

The response is bandpass. $\qquad\qquad\qquad\qquad\qquad f_0 = 397.887$

105

(c) For this part of the exercise, we want to use Electronics Workbench rather than do the approximations. The circuit and results follow:

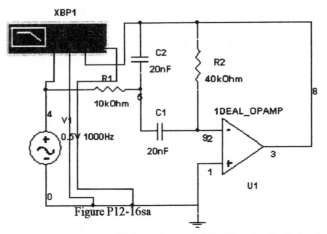

Figure P12-16sa

Notice that for this simulation we have used another instrument in Electronic Workbench - the bode plotter. Double clicking on the symbol brings up the following:

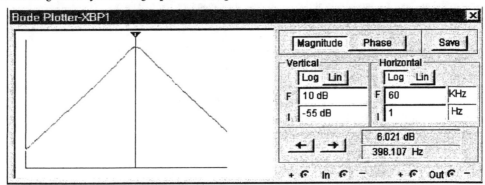

Figure P12-16sb

The center frequency is chosen by sliding the center marker to the desired location. The readout is 398.107 Hz. Very close to the calculated 397.887 Hz.

(d) The exact values of the magnitude of the output for a 0.5 V input at $\omega = 0.5\omega_0$, ω_0, and $2\omega_0$ are

$$0.5 \left| T_V(0.5 \cdot j \cdot \omega_0) \right| = 0.555$$

$$0.5 \left| T_V(j \cdot \omega_0) \right| = 1$$

$$0.5 \left| T_V(2 \cdot j \cdot \omega_0) \right| = 0.555$$

We can "connect" AC Voltmeter to the output and set the input frequency and voltage of the AC source to get the simulator results at $\omega = 0.5\omega_0$, ω_0, and $2\omega_0$. They are .560, .995 and .548 respectively.

106

12-25 A parallel RLC circuit with the following values of resistance, center frequency and bandwidth is required:

$$R := 40 \cdot k\Omega \qquad f_0 := 100 \cdot MHz \qquad B := 2 \cdot \pi \cdot 100 \cdot kHz$$

Calculate L and C to achieve the design.

Solution: We set up a solve block using MathCad with the equations for bandwidth and resonance frequency to solve for L and C. The guesses are

$$C := 1 \cdot pF \qquad L := 1\mu H$$

Given

$$B = \frac{1}{R \cdot C} \qquad f_0 = \frac{1}{2 \cdot \pi \cdot \sqrt{L \cdot C}}$$

$$\begin{pmatrix} C \\ L \end{pmatrix} := Find(C, L) \qquad C = 39.789 \, pF \qquad nH := 10^{-9} \cdot H$$

$$L = 63.662 \, nH$$

Note that we have established the units of nH so that we can get more significant figures for the value of L. We now simulate the circuit in Orcad Capture (Figure 12-25sa).

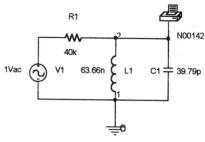

Figure 12-25sa

Once Probe is run (Figure 12-25sb) the Probe Cursor function allows one to compute the bandwidth (99.805 kHz) and to find the center frequency, f_0 (100.001 MHz).

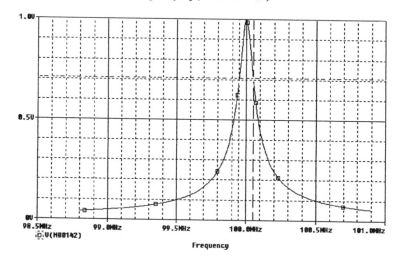

Probe Cursor used to find BW.

Probe Cursor used to find f_0.

Figure 12-25sb

107

12-27 (D) A series RLC circuit is to be used as a notch filter to eliminate a bothersome 60-Hz hum in an audio channel. The signal source has a Thevenin resistance of 600 Ω. Select values of L and C so the upper cutoff frequency is below 200 Hz.

Solution: The given values are (we have selected ω_{C2} based on the problem statement):

$$R := 600 \cdot \Omega \qquad f_0 := 60 \cdot Hz \qquad \omega_0 := 2 \cdot \pi \cdot f_0 \qquad \omega_{C2} := 2 \cdot \pi \cdot 150 \cdot Hz$$

The following solve block will yield the desired values of L and C:

$$L := 1 \cdot H \qquad\qquad C := 10 \mu F$$

Given

$$\omega_{C2} = \frac{R}{2 \cdot L} + \sqrt{\left(\frac{R}{2 \cdot L}\right)^2 + \frac{1}{L \cdot C}} \qquad \omega_0 = \frac{1}{\sqrt{L \cdot C}}$$

$$\binom{L}{C} := Find(L, C) \qquad\qquad L = 0.758\,H$$

$$C = 9.284\,\mu F$$

We now simulate the circuit in Electronics Workbench. The schematic (Figure 12-27sa) and results (Figure 12-27sb) follow: The markers and grids are buttons on the toolbar of the Analysis Graphs window. The markers can be positioned by the mouse cursor and are set so the amplitude is 707 mV (or as close as the resolution will allow). In this case, the upper cutoff frequency is ~ 150 Hz as designed.

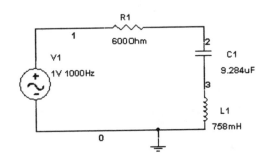

Figure 12-27sa

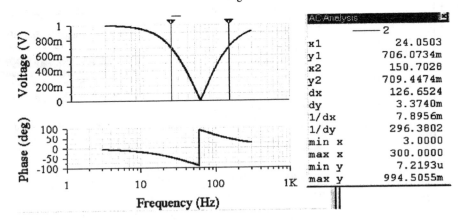

Figure 12-27sb

12-48 Construct a 1st order low-pass transfer function with a dc gain of 10, a bandwidth less than 250 rad/s, and a step response that rises to 50% of its final value in less than 4 ms.

Solution: The transfer function for a 1st order low-pass filter is:

$$T(s) = \frac{K}{\dfrac{s}{\omega_C} + 1}$$

where the DC gain value is given by K and the bandwidth is given by ω_C. We have the following for the step response.

$$\frac{10}{s \cdot \left(\dfrac{s}{\omega_C} + 1 \right)}$$

has inverse Laplace transform

$$10 \cdot \omega_C \cdot \left(\frac{1}{\omega_C} - \frac{1}{\omega_C} \cdot \exp\left(-\omega_C \cdot t \right) \right)$$

simplifies to

$$10 - 10 \cdot \exp\left(-\omega_C \cdot t \right)$$

For the step response to rise to 50% of its final value in less than 4 ms, we need

$$10 - 10 \cdot \exp\left(-\omega_C \cdot 4 \cdot 10^{-3} \right) = 5$$

has solution(s)

$$\omega_C := 250 \cdot \ln(2) \qquad\qquad \omega_C = 173.287$$

So we need $173.287 < \omega_C < 250$. We choose

$$\omega_C := 200 s^{-1}$$

Since we want to simulate this in Orcad Capture, we will design a first order active lowpass filter. Using the non-inverting amplifier with a gain of 10 and C = 1 μF,

$$C := 1 \cdot \mu F \qquad\qquad R := \frac{1}{\omega_C \cdot C} \qquad\qquad R = 5 \times 10^3 \, \Omega$$

We begin the analysis with a transient response to see if the rise time criterion has been satisfied . The circuit (Figure P12-48sa) shows a pulse voltage source with 1 volt maximum. The transient repines is selected under the Analysis Setup. The response is graphed (Figure P12-48sc) and the 50 % point is measured with the cursor. It is 3.5108 ms, less than 4 ms required. The frequency response circuit is shown in Figure P12-48sb. Its response is shown in Figure P12-48sd.

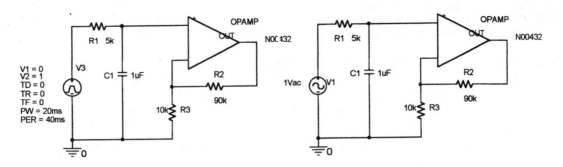

Figure P12-48sa Figure P12-48sb

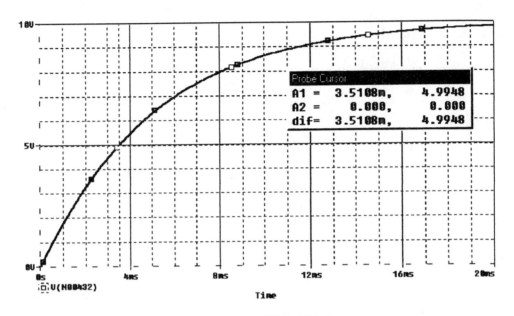

Figure P12-48sc

Next we do the AC analysis to check the cutoff frequency. The only change to the circuit is to substitute a 1 V AC source for the pulse source, set up the AC analysis and turn off the transient analysis. Figure P12-48sd shows the response and cutoff frequency of

$$\omega_C := 2 \cdot \pi \cdot 31.750 \qquad \omega_C = 199.491$$

This is less than the 250 rad/s maximum required by the task.

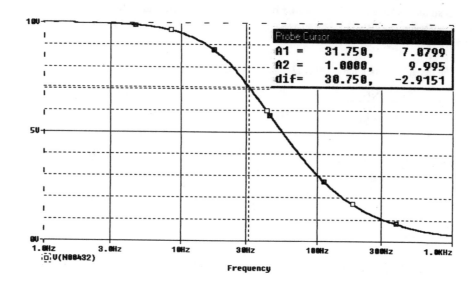

Figure P12-48sc

12-50 Construct a 2nd order bandpass transfer function with a midband gain of 10, a center frequency of 100 kHz, a bandwidth less than 50 kHz, and a step response that decays to less that 20 % of its peak value in less than 5 μs.

Solution: The transfer function for a 2nd order bandpass filter is

$$T(s) = \frac{K \cdot s}{s^2 + 2 \cdot \zeta \cdot \omega_0 \cdot s + \omega_0^2}$$

where, for this design

$$T_{max} = \frac{K}{2 \cdot \zeta \cdot \omega_0} = 10 \qquad B = 2 \cdot \zeta_{max} \cdot \omega_0 = 2 \cdot \pi \cdot 50 \cdot 10^3 \qquad \omega_0 := 2 \cdot \pi \cdot 10^5$$

so that $\qquad \zeta_{max} := \dfrac{\pi \cdot 50 \cdot 10^3}{\omega_0} \qquad\qquad \zeta_{max} = 0.25$

The step response is

$$\frac{K}{s^2 + 2 \cdot \zeta \cdot \omega_0 \cdot s + \omega_0^2}$$

has inverse Laplace transform

$$-K \cdot \frac{\exp(-\zeta \cdot \omega_0 \cdot t)}{\left[\omega_0^2 \cdot (\zeta - 1) \cdot (\zeta + 1)\right]} \cdot \left(\omega_0^2 - \zeta^2 \cdot \omega_0^2\right)^{\left(\frac{1}{2}\right)} \cdot \sin\left[\left[-\omega_0^2 \cdot (\zeta - 1) \cdot (\zeta + 1)\right]^{\left(\frac{1}{2}\right)} \cdot t\right]$$

The step response decay criterion implies

$$\exp\left(-\zeta_{min} \cdot \omega_0 \cdot 5 \cdot 10^{-6}\right) = 0.2$$

has solution(s)

$$\zeta_{min} := \frac{321887.58248682007492}{\omega_0} \qquad\qquad \zeta_{min} = 0.512$$

Obviously, there is no value for ζ that satisfies the criteria. Therefore, there is no solution.

12-54 (A) HIGH-FREQUENCY MODEL OF A RESISTOR

Figure P12-54 shows a circuit model of a resistor R that includes parasitic capacitance C and parasitic lead inductance L. The purpose of this problem is to investigate the effect of these parasitic elements on the high-frequency characteristics of the resistor device.

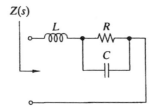

Figure P12-54

(a) Derive an expression for the impedance Z(s) of the circuit model in Figure P12-54 in terms of the circuit parameters R, L, and C.

(b) For R = 10 kΩ, L = 2 μH, and C = 4 pF find the poles and zeros of $Z(s)$. Construct a Bode plot of the straight-line asymptotes of $|Z(j\omega)|$. Do not express $|Z(j\omega)|$ in dB. Use a logarithmic scale with the impedance magnitude expressed in ohms. Identify the corner frequencies and slopes (in ohms/decade) in this plot.

(c) Use the straight-line Bode plot in (b) to identify the frequency range (in Hz) over which the resistor device appears to be: (1) A 10-kΩ resistance, (2) a 4-pF capacitance, and (3) a 2-μH inductor.

(d) Over what frequency range can you safely use this device in a circuit design and treat it as a pure 10-kΩ resistor?

Solution:

(a) The expression is

$$Z(s) = Ls + \frac{1}{Cs + \dfrac{1}{R}} = \frac{RLCs^2 + Ls + R}{RCs + 1}$$

(b) We will use the plotting capability of MATLAB to plot the exact function for this part of the problem. The m-file and plot (Figure P12-54sa) follow:

```
%  Component values are
R=10000;
L=2*10^-6;
C=4*10^-12;
% The range for f is set.
f=10^5:2*10^5:10^9;
w=2*pi*f;
s=j*w;
% Now generate the exact impedance plot.
Z=(R*L*C*s.^2+L*s+R)./(R*C*s+1);
Zmag=abs(Z);
semilogx(f,Zmag)
xlabel('Frequency, Hz')
ylabel('|Z|, ohms')
% Do the point where there is a transision between R and C
text((2*pi*R*C)^-1,10^4/sqrt(2),'o')
%
1/2/pi/R/C
1/2/pi/sqrt(L*C)
%
```

113

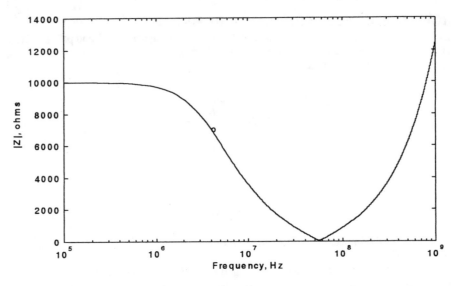

Figure P12-54sa

(c) The transition frequency between resistance (slope = 0) and capacitance (slope = -1) is 3.979 MHz (labeled with a "o" on the plot) and that between capacitance and inductance (slope = 1) is 56.27 MHz (the zero value on the plot).

(d) It acts like a resistor for frequencies less than 3.979 MHz.

13-12 The four terms in the Fourier series of a periodic signal are

$$v(t) = 25 \left[\sin(100\pi t) - \frac{1}{9}\sin(300\pi t) + \frac{1}{25}\sin(500\pi t) - \frac{1}{49}\sin(700\pi t) + ... \right]$$

(a) Find the period and fundamental frequency in rad/s and Hz.
(b) Does the waveform have even or odd symmetry?
(c) Plot the waveform using the truncated series above to confirm your answer in (b). Can you identify the waveform from this plot?

Solution: (a) $\omega = 100\pi$ rad/s, $f = 50$ Hz, and $T = 1/50$ s.
(b) The waveform has odd and halfwave symmetry since only odd harmonic sine terms are present.
(c) Using Excel to do the plot, we get the following graph:

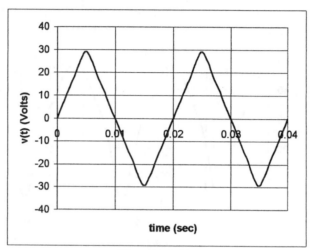

Figure P13-12s

The waveform is a triangular wave passing through the origin at t = 0.

115

13-13 The four terms in the Fourier series of a periodic signal are

$$v(t) = 30 + 20 \left[\cos(400\,t) - \frac{1}{5}\cos(800\,t) + \frac{1}{7}\cos(1200\,t) - \frac{1}{21}\cos(1600\,t) + \ldots \right]$$

(a) Find the period and fundamental frequency in rad/s and Hz.
(b) Does the waveform have even or odd symmetry?
(c) Plot the waveform using the truncated series above to confirm your answer in (b). Can you identify the waveform from this plot?

Solution: (a) By inspection, $\omega = 400$, $f = 200/\pi$ and $T = \pi/100$.
(b) The signal has even symmetry since only cosine terms are present.
(c) Using Excel to do the plot, we get the following graph:
(d)

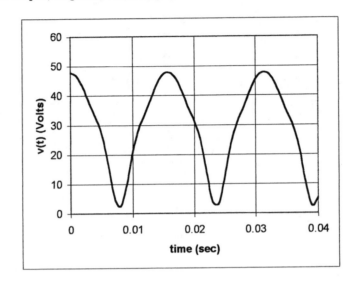

The signal appears to be a full wave rectified cosine.

13-29 Find the rms value of the periodic waveform in Figure P13-29 and write an expression for the total average power the waveform delivers to a resistor R. Find the Fourier coefficients of the waveform. What fraction of the total average power is carried by dc component plus the first three nonzero ac components in the Fourier series of the waveform?

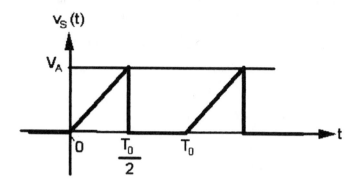

Figure P13-29

Solution: The m-file below is used for the rms value, power delivered to R, and the Fourier coefficients.

```
% Define the symbols.
syms T0 VA t R n
% Calculate Vrms and the power delivered to a load, R.
Vrms=sqrt(1/T0*int((VA*2*t/T0)^2,t,0,T0/2))
Ptotal=Vrms^2/R
% Calculate the Fourier Coefficients.
a0=1/T0*int(VA*2*t/T0,t,0,T0/2)
an=2/T0*int(VA*2*t/T0*cos(2*pi*n*t/T0),t,0,T0/2)
bn=2/T0*int(VA*2*t/T0*sin(2*pi*n*t/T0),t,0,T0/2)
%
```

The results are:

```
Vrms = 1/6*6^(1/2)*(VA^2)^(1/2)
Ptotal = 1/6*VA^2/R
a0 = 1/4*VA
an = VA*(cos(pi*n)+pi*n*sin(pi*n)-1)/pi^2/n^2
bn = -VA*(-sin(pi*n)+pi*n*cos(pi*n))/pi^2/n^2
```

Before we calculate the Vn, we will simplify the values for an and bn ($\sin(n\pi) = 0$). The m-file continues:

```
% Take the simplified results (sin(n*pi) = 0) from the command line.
an=VA*(cos(pi*n)-1)/pi^2/n^2;
bn=-VA*(pi*n*cos(pi*n))/pi^2/n^2;
Vn=sqrt(an^2+bn^2)
%
```

with the following results:

Vn = (VA^2*(cos(pi*n)+pi*n*sin(pi*n)-1)^2/pi^4/n^4+VA^2*(-sin(pi*n)+pi*n*cos(pi*n))^2/pi^4/n^4)^(1/2)

MATLAB does not do a very good job of simplifying this, so we do it here.

$$V_n = \frac{V_A}{n\pi}\sqrt{\left(\frac{\cos(n\pi)-1}{n\pi}\right)^2 + 1}$$

The fraction of the total average power carried by dc component plus the first three nonzero ac components in the Fourier series of the waveform is calculated in MATLAB. The continuation of the m-file and the result follow:

```
% Since we want a fraction of the power, we can assign values to the
waveform.
T0=1;
R=1;
VA=1;
% Now calculate the coefficients.
a0=VA/4;
n=1;
a1=VA*(cos(pi*n)-1)/pi^2/n^2;
b1=-VA*(pi*n*cos(pi*n))/pi^2/n^2;
V1=sqrt(a1^2+b1^2);
n=2;
a2=VA*(cos(pi*n)-1)/pi^2/n^2;
b2=-VA*(pi*n*cos(pi*n))/pi^2/n^2;
V2=sqrt(a2^2+b2^2);
n=3;
a3=VA*(cos(pi*n)-1)/pi^2/n^2;
b3=-VA*(pi*n*cos(pi*n))/pi^2/n^2;
V3=sqrt(a3^2+b3^2);
% Calculate the total power.
Ptotal=1/6*VA^2/R;
% Calculate the power in the dc and first 3 ac terms.
Ppartial=a0^2/R+(V1^2+V2^2+V3^2)/2*R;
% Calculate the percentage of the total power.
Percent=100*Ppartial/Ptotal
%
```

resulting in

```
Percent =   91.34408398557112
```

13-37 (A) STEADY-STATE RESPONSE FOR A PERIODIC IMPULSE TRAIN

A periodic impulse train can approximate a pulse train when the individual pulse durations are very short compared with the circuit response time. This example explores the response of a first-order circuit to a periodic impulse train. A linear circuit whose impulse response is

$$h(t) = \frac{1}{T_0} \cdot e^{\frac{-t}{T_0}} \cdot u(t)$$

is driven by a periodic impulse train

$$x(t) = T_0 \cdot \sum_{n=-\infty}^{\infty} \delta(t - n \cdot T_0)$$

(a) Use Eq. (13-3) to find the Fourier coefficients of $x(t)$. Write a general expression for the Fourier series of the input $x(t)$.

(b) Derive expressions for the Fourier coefficients of the steady-state output $y(t)$ when the input is the periodic impulse train $x(t)$.

(c) Write a general expression for the Fourier series of the statdy-state output $y(t)$.

(d) Use a computer tool to generate a plot of a truncated Fourier series that gives a reasonable approximation of the steady-state response found in (c).

Solution:

(a) The Fourier coefficients of $x(t)$ are

$$a_0 = \frac{1}{T_0} \cdot \lim_{x \to 0} \int_{-x}^{x} T_0 \cdot \delta(t) \, dt = 1 \qquad\qquad T_0 := 1$$

$$a_n = \frac{1}{T_0} \cdot \lim_{x \to 0} \int_{-x}^{x} T_0 \cdot \delta(t) \cos\left(2 \cdot \pi \cdot n \cdot \frac{t}{T_0} \right) dt = 1$$

$$b_n = \frac{1}{T_0} \cdot \lim_{x \to 0} \int_{-x}^{x} T_0 \cdot \delta(t) \sin\left(2 \cdot \pi \cdot n \cdot \frac{t}{T_0} \right) dt = 0$$

so that

$$x(t) = 1 + \sum_{n=1}^{\infty} \cos\left(2 \cdot \pi \cdot n \cdot \frac{t}{T_0} \right) \qquad \text{hence} \qquad A_n := 1 \qquad \phi_n := 0$$

Note that the subcript, n, is input with a "[" to denote an index. Do not use the notational subscript input with a ".". n is globally defined below.

(b) If $\quad h(t) = \frac{1}{T_0} \cdot e^{\frac{-t}{T_0}} \quad$ then $\quad \frac{1}{T_0} \cdot e^{\frac{-t}{T_0}}$

has Laplace transform

$$\frac{1}{\left[T_0 \cdot \left(s + \frac{1}{T_0} \right) \right]}$$

simplifies to

$$T(s, T_0) := \frac{1}{(s \cdot T_0 + 1)}$$

The amplitude and phase angle of the steady-state output for the periodic impulse train input are

$$B_n := A_n \cdot \left| T\left(\frac{j \cdot 2 \cdot \pi \cdot n}{T_0}, T_0\right) \right| \qquad \theta_n := \phi_n + \arg\left(T\left(\frac{j \cdot 2 \cdot \pi \cdot n}{T_0}, T_0\right) \right)$$

(c) The general expression for the Fourier series of the statdy-state output $y(t)$ is

$$y(t) = \sum_{n=0}^{\infty} |B_n| \cdot \cos\left[\left(2 \cdot \pi \cdot n \cdot \frac{t}{T_0}\right) + \theta_n\right]$$

(d) A truncated Fourier series that gives a reasonable approximation of the steady-state response found in (c) is

$$y(t) := \sum_{n=0}^{N} B_n \cdot \cos\left[\left(2 \cdot \pi \cdot n \cdot \frac{t}{T_0}\right) + \theta_n\right] \qquad t := 0, \frac{1}{10 \cdot N} .. 2 \qquad N \equiv 50 \qquad n \equiv 0 .. N$$

The globaly defined values of N and n are located here to enable us to see the change in the graph when we change N. The resolution of the graph also changes with N to capture the finer details as we increase N.

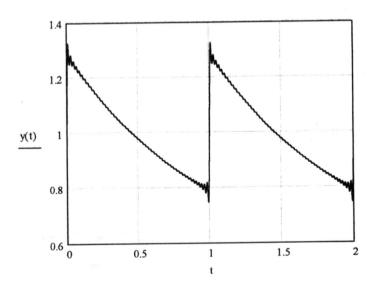

13-39 (D) FILTER DESIGN REQUIREMENTS

A periodic signal with odd symmetry is to be applied to inputs of a bank of second-order bandpass filters. Each filter must pass one specific harmonic with a gain of 0 dB and have a gain less than -20 dB at the two adjacent harmonic frequencies. What quality factor Q is required in each of the second-order filters?

Since the requirement is for all harmonics we will select an upper harmonic - the 7th, but it really does not matter which harmonic was selected as long as it was an odd harmonic.

We also need to select a period for our fundamental, e.g. $T_0 = 2 \cdot \pi \cdot 10^{-6}$ $\omega_0 := 10^6$

To be centered at the 7th harmonic our cutoff frequency is $\omega_C := 7 \cdot \omega_0$

and the transfer functiion is $T(s, Q) := \dfrac{\dfrac{7 \cdot \omega_0}{Q} \cdot s}{s^2 + \dfrac{7 \cdot \omega_0}{Q} \cdot s + \left(7 \cdot \omega_0\right)^2}$ then, since the waveform has odd symmetry

all harmonics are present and the two adjacent harmonics are the 6th and 8th. We will try various Q's and select the first one that is greater than 20 dB down from the adjacent harmonic.

$Q := 10, 20 .. 60$

$Q =$	$20 \cdot \log\left(\left\|T\left(j \cdot 6 \cdot \omega_0, Q\right)\right\|\right) =$	$20 \cdot \log\left(\left\|T\left(j \cdot 8 \cdot \omega_0, Q\right)\right\|\right) =$	
10	-10.245	-9.125	
20	-15.946	-14.727	
30	-19.406	-18.167	
40	-21.883	-20.637	<---Q = 40 is the minimum
50	-23.811	-22.562	
60	-25.389	-24.138	

14-11 (D) The high-pass straight-line gain response in Figure P14-11 provides a two-pole stopband roll off at low frequencies.

   (a) Construct a transfer function $T(s)$ that has this gain response using only real poles and zeros.

   (b) Design an active RC circuit to realize the $T(s)$ found in (a) using only one OP AMP. Scale the circuit so the element values are in practical ranges.

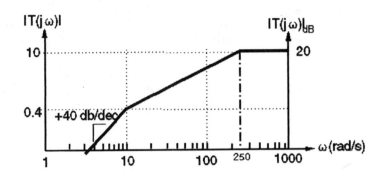

Figure P14-11

Solution: (a) The transfer function is straightforward:

$$T(s) = \frac{10s^2}{(s+10)(s+250)} = \left[\frac{s}{s+10}\right][10]\left[\frac{s}{s+250}\right] = T_1 \times T_2 \times T_3$$

(b) The cascaded transfer function are a high pass RC filter for T_1, an OP AMP stage with gain 20 for T_2, and a high pass RC filter for T_3. The MATLAB m-file for calculating the components and the results follow:

```
% For the first stage, the prototype components are
C1=1;
R1=1/10;
% Scaling these to get reasonable component values
k1=10^6;
C1=C1/k1
R1=R1*k1
% The resistors for the non-inverting OP AMP stage are
RB=10^4
RA=9*RB
% For the third stage, the prototype components are
C3=1;
R3=1/250;
% Scaling these to get reasonable component values
k3=10^6;
C3=C3/k3
R3=R3*k3

Results:

C1 = 1.000000000000000e-006
R1 = 100000
RB = 10000
RA = 90000
C3 = 1.000000000000000e-006
R3 = 4000
```

The schematic (drawn in Orcad Capture) and the results of the simulation are shown in Figures P14-11sa and P14-11sb respectively below:

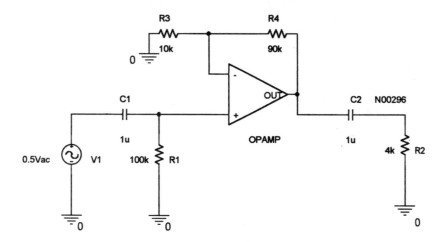

Figure P14-11sa

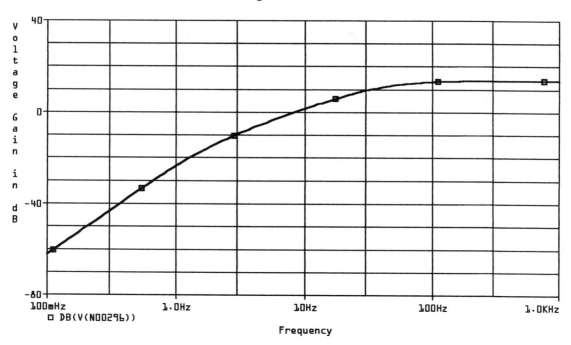

Frequency

Figure P14-11sb

14-12 The straight-line gain response in Figure P14-12 is the design specification for an audio preamplifier. Note that frequencies are specified in Hz.
(a) Construct a transfer function T(s) that has this gain response using only real poles and zeros.
(b) Design an active RC circuit to realize the *T(s)* found in (a) using only one OP AMP. Scale the circuit so the element values are in practical ranges.

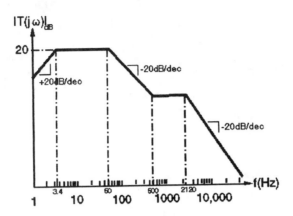

Figure P14-12

Solution: We will solve this problem using Mathcad. The transfer function is written down by inspection of the figure.

$$T(s) := \frac{\dfrac{10 \cdot s}{2 \cdot \pi \cdot 3.4} \left(\dfrac{s}{2 \cdot \pi \cdot 500} + 1 \right)}{\left(\dfrac{s}{2 \cdot \pi \cdot 3.4} + 1 \right) \cdot \left(\dfrac{s}{2 \cdot \pi \cdot 50} + 1 \right) \cdot \left(\dfrac{s}{2 \cdot \pi \cdot 2120} + 1 \right)}$$

Let's check the response to be sure.

$$f := 1, 2 .. 10000$$

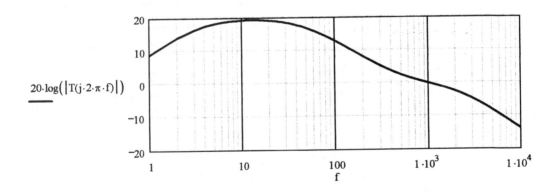

(b) To design the circuit we need to rearrange T(s).

$$T(s) = \frac{\dfrac{10 \cdot s}{2 \cdot \pi \cdot 3.4} \cdot \left(\dfrac{s}{2 \cdot \pi \cdot 500} + 1\right)}{\left(\dfrac{s}{2 \cdot \pi \cdot 3.4} + 1\right) \cdot \left(\dfrac{s}{2 \cdot \pi \cdot 50} + 1\right) \cdot \left(\dfrac{s}{2 \cdot \pi \cdot 2120} + 1\right)} = \frac{\dfrac{\left(\dfrac{s}{2 \cdot \pi \cdot 500} + 1\right)}{\left(\dfrac{s}{2 \cdot \pi \cdot 50} + 1\right) \cdot \left(\dfrac{s}{2 \cdot \pi \cdot 2120} + 1\right)}}{\dfrac{\dfrac{s}{2 \cdot \pi \cdot 3.4} + 1}{\dfrac{10s}{2 \cdot \pi \cdot 3.4}}}$$

We can do this design with an inverting amplifier with transfer function $T(s) = Z_2/Z_1$ where

$$Z_1 = \frac{\dfrac{s}{2 \cdot \pi \cdot 3.4} + 1}{\dfrac{10s}{2 \cdot \pi \cdot 3.4}} = 0.1 + \frac{2 \cdot \pi \cdot 0.34}{s} \qquad Z_2 = \frac{\left(\dfrac{s}{2 \cdot \pi \cdot 500} + 1\right)}{\left(\dfrac{s}{2 \cdot \pi \cdot 50} + 1\right) \cdot \left(\dfrac{s}{2 \cdot \pi \cdot 2120} + 1\right)}$$

Z_1 is a series RC circuit with scaling factor and values

$$k_1 := 10^5 \qquad R_1 := k_1 \cdot 0.1 \qquad R_1 = 1 \times 10^4$$

$$C_1 := \frac{1}{2 \cdot \pi \cdot 0.34 \cdot k_1} \qquad C_1 = 4.681 \times 10^{-6}$$

Z_2 needs some more rearranging.

$$Z_2 = \frac{\left(\dfrac{s}{2 \cdot \pi \cdot 500} + 1\right)}{\left(\dfrac{s}{2 \cdot \pi \cdot 50} + 1\right) \cdot \left(\dfrac{s}{2 \cdot \pi \cdot 2120} + 1\right)}$$

simplifies to   *adding the right hand side*

$$Z_2 = 424 \cdot (s + 1000 \cdot \pi) \cdot \frac{\pi}{[(s + 100 \cdot \pi) \cdot (s + 4240 \cdot \pi)]} = \frac{A}{s + 100 \cdot \pi} + \frac{B}{s + 4240 \cdot \pi}$$

simplifies to

$$Z_2 = 424 \cdot (s + 1000 \cdot \pi) \cdot \frac{\pi}{[(s + 100 \cdot \pi) \cdot (s + 4240 \cdot \pi)]} = \frac{(A \cdot s + 4240 \cdot A \cdot \pi + B \cdot s + 100 \cdot B \cdot \pi)}{[(s + 100 \cdot \pi) \cdot (s + 4240 \cdot \pi)]}$$

Now compare the left and right hand sides to get a symbolic solve block.

Given

$$A + B = 424 \cdot \pi$$

$$4240 A \cdot \pi + 100 \cdot B \cdot \pi = 424 \cdot 1000 \pi \cdot \pi$$

$$\text{Find}(A, B) \rightarrow \begin{pmatrix} \dfrac{2120}{23} \cdot \pi \\[2mm] \dfrac{7632}{23} \cdot \pi \end{pmatrix}$$

So that we have

$$Z_2 = \frac{2120}{23} \cdot \frac{\pi}{s + 100 \cdot \pi} + \frac{7632}{23} \cdot \frac{\pi}{s + 4240 \cdot \pi} = \frac{1}{\left(\dfrac{23 \cdot s}{2120 \cdot \pi} + \dfrac{2300}{2120}\right)} + \frac{1}{\left(\dfrac{23 \cdot s}{7632 \cdot \pi} + \dfrac{23 \cdot 4240}{7632}\right)}$$

$$Z_2 = \frac{1}{\left(\dfrac{23 \cdot s}{2120 \cdot \pi} + \dfrac{2300}{2120}\right)} + \frac{1}{\left(\dfrac{23 \cdot s}{7632 \cdot \pi} + \dfrac{23 \cdot 4240}{7632}\right)} = \frac{1}{C_2 \cdot s + \dfrac{1}{R_2}} + \frac{1}{C_3 \cdot s + \dfrac{1}{R_3}}$$

This is a series combination of parallel combinations of R_2 and C_2 and R_3 and C_3. Using the scaling factor $k_2 := 10^5$ we have

$$R_2 := k_2 \cdot \frac{2120}{2300} \qquad\qquad R_2 = 9.217 \times 10^4$$

$$C_2 := \frac{1}{k_2} \cdot \frac{23}{2120 \cdot \pi} \qquad\qquad C_2 = 3.453 \times 10^{-8}$$

$$R_3 := k_2 \cdot \frac{7632}{23 \cdot 4240} \qquad\qquad R_3 = 7.826 \times 10^3$$

$$C_3 := \frac{1}{k_2} \cdot \frac{23}{7632 \cdot \pi} \qquad\qquad C_3 = 9.593 \times 10^{-9}$$

We now go to Electronics Workbench to check our design. The circuit and simulation results follow:

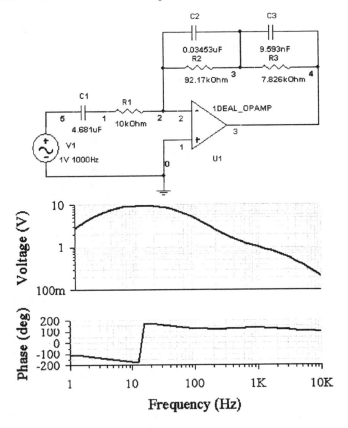

14-19 (D) Design second-order active RC circuits to meet the following requirements.

Problem Type	ω_0(rad/s)	ζ	Constraints
High Pass	2000	?	20 dB gain at corner frequency.

Solution: Use the equal R, equal C method. The transfer function is

$$T(s) = \frac{\mu(RC\,s)^2}{(RC\,s)^2 + (3 - \mu)RC\,s + 1}$$

The prototype circuit is shown in Figure P14-19sa

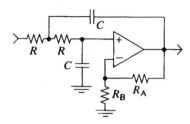

Figure P14-19sa

For the gain of 10 at the corner frequency, we need $\mu/(3-\mu) = 10$. The MATLAB command line ordinarily would be just fine for this and other subsequent calculations, but what if we wanted to be able to do a number of designs like this? It is wise to create an m-file. It and the results follow:

```
% Second order high pass filter using the equal element value method.
% Enter the gain and cutoff frequency.
K=input('What is the desired gain?  ')
wC=input('What is the desired cutoff frequency in rad/sec?  ')
% Calculate mu
mu=3*K/(1+K);
% Enter the value of RB
RB=input('Enter the value of RB (ohms).  ')
% Calculate RA
RA=(mu-1)*RB
% Enter the value of R
R=input('Enter the value of R (ohms).  ')
% Calculate C
C=1/R/wC
%
```

Results:

```
What is the desired gain?  10
K = 10
What is the desired cutoff frequency in rad/sec?  2000
wC = 2000
Enter the value of RB (ohms).  10000
RB = 10000
RA = 1.727272727272727e+004
Enter the value of R (ohms).  10000
R = 10000
C = 5.000000000000000e-008
```

We now simulate the circuit using Electronics Workbench. The schematic and response are shown in Figures Figure P14-19sb and Figure P14-19sc.

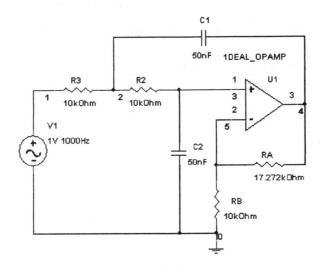

Figure P14-19sb

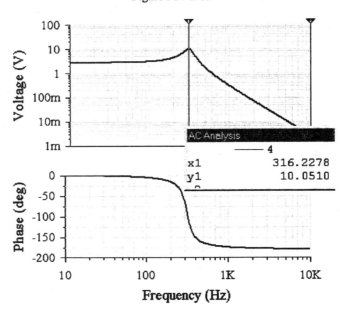

Figure P14-19sc

The cutoff frequency (316.23 Hz) and gain (10.05) are very close to the design values of 318 Hz and 10. Of course, the accuracy of the graph display depends on the resolution (number of frequency points/decade for this case is 100).

14-21 (D) Design second-order active RC circuits to meet the following requirements.

Problem Type	ω_0(rad/s)	ζ	Constraints
High Pass	1000	0.75	High-frequency gain of 40 dB.

Solution: We use a similar m-file to that generated for problem 14-19. This time, the unity gain method is used. It and the results follow.

```
% Second order high pass filter using the unity gain method
% with the damping ratio given.
% Enter the gain, the damping ratio and cutoff frequency.
K=input('What is the desired gain?  ')
wC=input('What is the desired cutoff frequency in rad/sec?  ')
zeta=input('What is the desired damping ratio?  ')
% Enter the value of RB
RB=input('Enter the value of RB (ohms)for the gain stage.  ')
% Calculate RA
RA=(K-1)*RB
% Enter the value of R1
R1=input('Enter the value of R1 (ohms).  ')
% Calculate R2 and C.
R2=R1/zeta^2
C=1/wC/sqrt(R1*R2)
%
```

The results are:

```
What is the desired gain?  100
K = 100
What is the desired cutoff frequency in rad/sec?  1000
wC = 1000
What is the desired damping ratio?  .75
zeta = 0.75000000000000
Enter the value of RB (ohms)for the gain stage.  10000
RB = 10000
RA = 990000
Enter the value of R1 (ohmns).  10000
R1 = 10000
R2 = 1.777777777777778e+004
C = 7.500000000000001e-008
```

Now go to Electronics Workbench the check the design. The circuit, Figure P14-21sa, and results, Figure P14-21sb, follow:

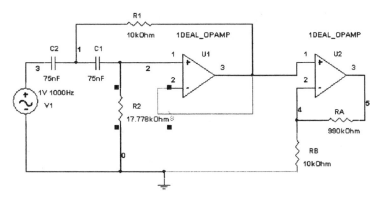

Figure P14-21sa

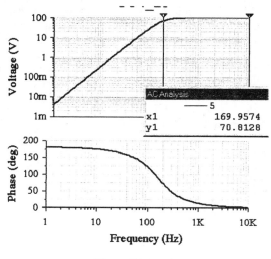

Figure P14-21sb

The cutoff frequency of 169.95 Hz is very close to the design value of 159 Hz. Of course, the accuracy of the graph display depends on the resolution (number of frequency points/decade for this case is 25).

14-33 Construct a transfer function with a fourth-order Chebychev lowpass response with $\omega_C = 500$ rad/s and a passband gain of 0 dB. Find the gain (in dB) at $2\omega_C$, $5\omega_C$, and $10\omega_C$. Plot the straight-line gain response on the range $0.1\,\omega_C \leq \omega_C \leq 10\omega_C$ and sketch the actual response.

Solution: The normalized fourth-order Chebychev lowpass response transfer function is taken from Table 14-2. We use MATLAB to complete the problem. (Since we are using a computer tool, we will add the straight-line gain response by hand.) The response curve is shown in Figure P14-33s.

```
% Design parameters
wC=500;
K=1;
% Range of w for the plot
w=0.1*wC:0.01*wC:10*wC;
s=j*w;
% Chebychev lowpass response
T=1/sqrt(2)./((s/0.9502/wC).^2+0.1789*s/0.9502/wC+1)./((s/0.4425/wC).^2+0.927
6*s/0.4425/wC+1);
% Plot the response
semilogx(w,20*log(abs(T)))
% Make the plot pretty
xlabel('Frequency (rad/s)')
ylabel('T(s) (dB)')
grid
%
```

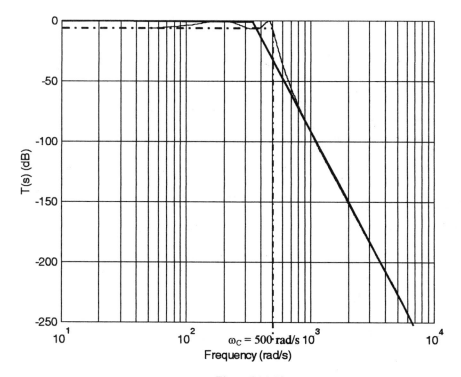

Figure P14-33s

14-34 (D) A low-pass filter specification requires $\omega_C = 2$ krad/s, a passband gain of 0 dB, and a stopband gain less than -30 dB at 10 krad/s.

(a) Find the lowest order (n) of a first-order cascade response that meets these requirements.
(b) Repeat (a) using a Butterworth response and a Chebychev response.
(c) Calculate the actual stopband gain at $\omega = 40$ krad/s for the responses found in (a) and (b).
(d) Which of these responses would you use to minimize the number of stages in a cascade design?

Solution: Use MATLAB for this problem. The m-file follows:

```
% Input of the design criteria.
wC=input('What is the desired cutoff frequency in rad/s?  ')
K=input('What is the desired passband gain?  ')
wmin=input('What is the stopband frequency criterion in rad/s?  ')
n=input('Estimate the value of n.  ')
s=j*wmin;
% First order cascade response.
Cascade=20*log10(abs(K/(s*sqrt(2^(1/n)-1)/wC+1)^n))
% Butterworth response
Butterworth=20*log10(K/sqrt(1+(wmin/wC)^(2*n)))
% Chebychev response
Chebychev=20*log10(K/sqrt(2)/sqrt(1+cosh(n*acosh(wmin/wC))))^2)
%
```

The results follow:

```
» P14_34
What is the desired cutoff frequency in rad/s?  2000
wC =   2000
What is the desired passband gain?  1
K =   1
What is the stopband frequency criterion in rad/s?  10000
wmin = 10000
Estimate the value of n.  1
n = 1
Cascade = -14.14973347970818
Butterworth = -14.14973347970818
Chebychev = -18.57332496431268
» P14_34
What is the desired cutoff frequency in rad/s?  2000
wC = 2000
What is the desired passband gain?  1
K = 1
What is the stopband frequency criterion in rad/s?  10000
wmin = 10000
Estimate the value of n.  2
n = 2
Cascade = -21.10400212716352
Butterworth = -27.96574333210429
Chebychev = -36.98970004336019
» P14_34
What is the desired cutoff frequency in rad/s?  2000
wC = 2000
What is the desired passband gain?  1
K = 1
```

```
What is the stopband frequency criterion in rad/s?  10000
wmin = 10000
Estimate the value of n.   3
n = 3
Cascade = -26.24840869100340
Butterworth = -41.93847819973556
Chebychev = -56.74302534188567
» P14_34
What is the desired cutoff frequency in rad/s?  2000
wC =   2000
What is the desired passband gain?   1
K = 1
What is the stopband frequency criterion in rad/s?  10000
wmin = 10000
Estimate the value of n.   4
n = 4
Cascade = -30.32672413846179
Butterworth = -55.91761146480600
Chebychev = -76.63874307105998
»
```

(a) and (b) In summary, the cascade response requires n = 4, the Butterworth requires n = 3 and the Chebychev requires n = 2.

(b) The m-file continues:

```
% Calculate the actual response for the result of Part I.
%
w=40000;
% First order cascade response.
n=4
Cascade=20*log10(abs(K/(s*sqrt(2^(1/n)-1)/wC+1)^n))
% Butterworth response
n=3
Butterworth=20*log10(K/sqrt(1+(wmin/wC)^(2*n)))
% Chebychev response
n=2
Chebychev=20*log10(K/sqrt(2)/sqrt(1+cosh(n*acosh(wmin/wC)))^2)
%
```

with the following results:

```
n = 4
Cascade = -30.32672413846179
n = 3
Butterworth = -41.93847819973556
n = 2
Chebychev = -36.98970004336019
```

(d) Use Chebychev to minimize the number of stages.

14-46 (D) A bandpass filter specification requires cutoff frequencies at 100 rad/s and 2 krad/s, a passband gain of 0 dB, and stopband gains less than -20 dB at 25 rad/s and 8 krad/s.
(a) Construct a bandpass transfer function $T(s)$ that meets the specification using Butterworth poles.
(b) Repeat (a) using Chebychev poles.
(c) Calculate the stopband gain at 25 rad/s and 4 krad/s for the transfer functions found in (a).
(d) Which one of these responses would you choose and why?

Solution: The bandpass requirements are:

$$T_{MAX} := 1 \quad \omega_{C1} := 100 \quad \omega_{C2} := 2000 \quad T_{MIN} := 0.1 \quad \omega_{MIN1} := 25 \quad \omega_{MIN2} := 8000$$

(a) Butterworth low pass requirement   Butterworth high pass requirement

$$\frac{1}{2} \frac{\ln\left[\left(\dfrac{T_{MAX}}{T_{MIN}}\right)^2 - 1\right]}{\ln\left(\dfrac{\omega_{MIN2}}{\omega_{C2}}\right)} = 1.657 \quad \text{or} \quad n = 2 \qquad \frac{1}{2} \frac{\ln\left[\left(\dfrac{T_{MAX}}{T_{MIN}}\right)^2 - 1\right]}{\ln\left(\dfrac{\omega_{C1}}{\omega_{MIN1}}\right)} = 1.657 \quad \text{or} \quad n = 2$$

so that

$$T_B(s) := \frac{1}{\left(\dfrac{s}{\omega_{C2}}\right)^2 + 1.414 \cdot \dfrac{s}{\omega_{C2}} + 1} \cdot \frac{s^2}{s^2 + 1.414 \cdot \omega_{C1} \cdot s + \omega_{C1}^2}$$

(b) Chebychev low pass requirement   Chebychev high pass requirement

$$\frac{\operatorname{acosh}\left[\sqrt{\left(\dfrac{T_{MAX}}{T_{MIN}}\right)^2 - 1}\right]}{\operatorname{acosh}\left(\dfrac{\omega_{MIN2}}{\omega_{C2}}\right)} = 1.448 \quad \text{or} \quad n = 2 \qquad \frac{\operatorname{acosh}\left[\sqrt{\left(\dfrac{T_{MAX}}{T_{MIN}}\right)^2 - 1}\right]}{\operatorname{acosh}\left(\dfrac{\omega_{C1}}{\omega_{MIN1}}\right)} = 1.448 \quad \text{or} \quad n = 2$$

so that

$$T_C(s) := \frac{\dfrac{1}{\sqrt{2}}}{\left(\dfrac{s}{0.8409 \cdot \omega_{C2}}\right)^2 + 0.7654 \cdot \left(\dfrac{s}{0.8409 \cdot \omega_{C2}}\right) + 1} \cdot \frac{s^2}{s^2 + 0.7654 \cdot \dfrac{\omega_{C1}}{0.8409} \cdot s + \left(\dfrac{\omega_{C1}}{0.8409}\right)^2}$$

(c) At 25 rad/s and 4 krad/s we have

$$20 \cdot \log\left(\left|T_B(25j)\right|\right) = -24.099 \qquad\qquad 20 \cdot \log\left(\left|T_B(4000j)\right|\right) = -12.304$$

$$20 \cdot \log\left(\left|T_C(25j)\right|\right) = -29.83 \qquad\qquad 20 \cdot \log\left(\left|T_C(4000j)\right|\right) = -16.984$$

Let's look at a graph of these responses. $\omega := 10, 12 .. 10000$

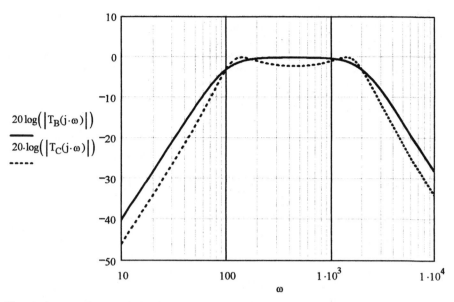

$$\frac{20\log\left(\left|T_B(j\cdot\omega)\right|\right)}{20\cdot\log\left(\left|T_C(j\cdot\omega)\right|\right)}$$

(d) Since both filter designs require essentially the same number of stages and components the answer to which filter is best can be simply summed up by - "It depends." If a flat response is needed in the pass band, the Butterworth is to one to use. If a sharper cutoff near the passband is needed in the stop bands and you can live the dip in the pass band, then choose the Chebychev.

14-52 (D) MODIFYING AN EXISTING DESIGN

An existing digital data channel uses a second-order Butterworth low-pass filter with a cutoff frequency of 10 kHz. Field experience reveals that noise is causing the equipment performance to fall below advertised levels. Analysis by the engineering department suggests that decreasing the filter gain by 20 dB at 30 kHz will solve the problem, provided the cutoff frequency does not change. The manufacturing department reports that adding anything more than one second-order stage will cause a major redesign and an unacceptable slip in scheduled deliveries. Design a second-order low-pass filter to connect in cascade with the existing second-order Butterworth filter. The frequency response of the circuit consisting of your second-order circuit in cascade with the original second-order Butterworth circuit must have a cutoff frequency of 10 kHz and reduce the gain at 30 kHz by at least 20 dB.

Solution: The existing filter has the following response:

$$f_C := 10 \cdot 10^3 \qquad\qquad T_1(s) := \dfrac{1}{\left(\dfrac{s}{2 \cdot \pi \cdot f_C}\right)^2 + \sqrt{2} \cdot \dfrac{s}{2 \cdot \pi f_C} + 1}$$

The added filter has the response:

$$T_2(s, \zeta, \omega_0) := \dfrac{1}{\left(\dfrac{s}{\omega_0}\right)^2 + 2 \cdot \zeta \cdot \dfrac{s}{\omega_0} + 1}$$

The added filter must have a gain of 1 at f_C and a gain of 0.1 at 30 kHz. The design requirements can be used to generate a solve block with guesses:

$$\omega_0 := 10^4 \qquad \zeta := .5$$

Given

$$\left| T_2 \left[(j \cdot 2 \cdot \pi \cdot f_C), \zeta, \omega_0 \right] \right| = 1 \qquad \left| T_2 \left[(j \cdot 2 \cdot \pi \cdot 3 \cdot 10^4), \zeta, \omega_0 \right] \right| = 0.1$$

$$\begin{pmatrix} \omega_0 \\ \zeta \end{pmatrix} := \text{Find}(\omega_0, \zeta) \qquad\qquad \begin{pmatrix} \omega_0 \\ \zeta \end{pmatrix} = \begin{pmatrix} 5.802 \times 10^4 \\ 0.455 \end{pmatrix}$$

Check the responses with a plot. $\omega := 1000, 1100 .. 200000$

$$\text{Gain}_1(\omega) := 20 \cdot \log\left(\left|T_1(j \cdot \omega)\right|\right)$$

$$\text{Gain}_{12}(\omega) := 20 \cdot \log\left(\left|T_1(j \cdot \omega)\right| \cdot \left|T_2(j \cdot \omega, \zeta, \omega_0)\right|\right)$$

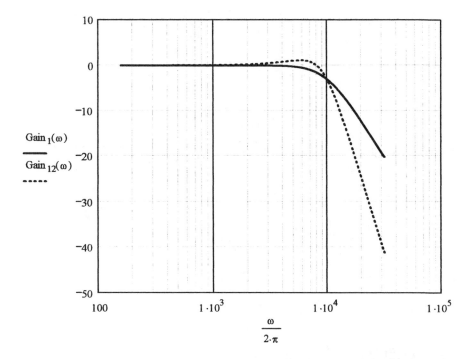

We use the trace feature by right clicking on the graph and selecting trace. A dialog box pops up with an option to "Track Data Points". Checking that box allows you to follow the graph lines to determine x, y values and also to copy and paste them. Doing this, I get that both lines cross the -2.9998 dB point at 9994.9 Hz. The existing filter is -19.302 dB at 30287 Hz. The cascaded filter is -39.476 at the same frequency - a difference of 20.174 dB. Of course, the exact numbers depend on the closeness of the data points.

15-1 Traditional, 16-1 Laplace Early. The input to the coupled coils in Figure P15-1 is a voltage source $v_S(t) = 10[\sin 2000t]$ V. The output is connected to an open circuit.

(a) Write the i-v relationships for the coupled inductors using the reference marks in the figure.

(b) Use the results in (a) and the specified input-output connections to solve for $v_2(t)$.

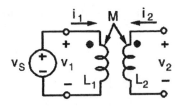

$$L_1=10 \text{ mH } L_2=5 \text{ mH}$$
$$M=7 \text{ mH}$$

Figure P15-1

Solution: (a) The i-v relationships for the coupled inductors are

$$v_1(t) = L_1 \frac{di_1(t)}{dt} + M \frac{di_2(t)}{dt} = 10 \sin 2000t$$

$$v_2(t) = L_2 \frac{di_2(t)}{dt} + M \frac{di_1(t)}{dt}$$

(b) Since $i_2(t)$ is 0 (open circuit), we have

$$L_1 \frac{di_1(t)}{dt} = 10 \sin 2000t \quad \text{hence} \quad \frac{di_1(t)}{dt} = \frac{10}{L_1} \sin 2000t$$

$$v_2(t) = M \frac{di_1(t)}{dt} = M \frac{10}{L_1} \sin 2000t = 7 \sin 2000t \ V$$

Let's go to a circuit simulator and check this out. Orcad Capture was used for this simulation. The schematic (Figure P15-1s) shows a 1-Ω resistor in series with the source and a 10 GΩ resistor at the output. These are needed to avoid infinite current and floating nodes respectively. The parameters of the coupled coils are the inductances and the coupling coefficient, k. Recall that $k = M/\sqrt{L_1 L_2}$. It appears the Electronics Workbench does not have a coupled coil in its parts list.

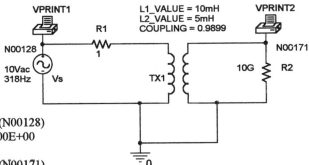

FREQ   VM(N00128) VP(N00128)
3.180E+02   1.000E+01   0.000E+00

FREQ   VM(N00171) VP(N00171)
3.180E+02   6.991E+00   2.865E+00

Figure P15-1s

The output is found to be: $v_2(t) = 6.99 \sin(2000t + 2.87°) V$ quite close to theory.

15-23 Traditional, 16-23 Laplace First. The input voltage in Figure P15-23 is a sinusoid $v_S(t) = 20 \cos 20000t$. Find the steady-state response of $i(t)$.

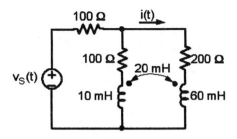

Figure P15-23

Solution: Define the center and right vertical branches to have currents I1 and I2 respectively. The voltages across the 10 mH and 60 mH inductors are V1 and V2 respectively. Two mesh equations (left loop and outside loop) along with the coupled coil constraints give four equations and four unknowns. These are solved in MATLAB to give the unknowns. The m-file and results follow:

```
% Define the components.
R1=100;
R2=100;
R3=200;
L1=0.01;
L2=0.06;
M=0.02;
VS=20;
w=20000;
% Calculate reactances.
X1=w*L1;
X2=w*L2;
XM=w*M;
%  Define symbols.
syms V1 V2 I1 I2
%  Two loop equations.
f1=-20+R1*(I1+I2)+R2*I1+V1;
f2=-20+R1*(I1+I2)+R3*I2+V2;
%  Coupled coil constraints
f3=j*X1*I1+j*XM*I2-V1;
f4=j*XM*I1+j*X2*I2-V2;
%  Solution
[V1 V2 I1 I2]=solve(f1, f2, f3, f4);
i=double(I2)
imag=abs(i)
iarg=angle(i)*180/pi
%
```

which results in

```
i = -0.01906693711968 - 0.00649087221095i
imag =  0.02014148736277
iarg =  -1.612001148413473e+002
```

We now go to Orcad Capture to simulate the circuit. The schematic (Figure 15-23s) and results follow:

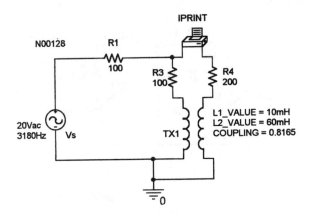

Figure 15-23s

FREQ   IM(V_PRINT3) IP(V_PRINT3)
3.180E+03   2.015E-02   -1.611E+02

The Probe results are in very close agreement to the MATLAB results found above.
That is, $i_L(t) = 20.15 \cos (20000t - 161.1°)$ mA

15-26 Traditional, 16-26 Laplace Early. The frequency of the voltage source in Figure P15-26 is adjustable. Find the frequency at which the impedance seen at the input interface is purely real.

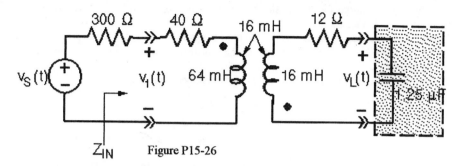

Figure P15-26

Solution: First, define the components.

$$R_1 := 300 \qquad L_1 := .064 \qquad C_L := 1.25 \cdot 10^{-6}$$

$$R_2 := 40 \qquad L_2 := .016$$

$$R_3 := 12 \qquad M := .016$$

The impedances, turns ratio and the coupling coefficient are

$$Z_1(\omega) := j \cdot \omega \cdot L_1 \qquad Z_2(\omega) := j \cdot \omega \cdot L_2 \qquad Z_L(\omega) := \frac{1}{j \cdot \omega \cdot C_L} \qquad k := \frac{M}{\sqrt{L_1 \cdot L_2}} \qquad n := \frac{M}{L_1}$$

The impedance reflected to the primary side is

$$Z_R(\omega) := \frac{Z_2(\omega) \cdot \left(1 - k^2\right) + Z_L(\omega) + R_3}{n^2}$$

The input impedance is then

$$Z_{in}(\omega) := R_2 + \left(Z_1(\omega)^{-1} + Z_R(\omega)^{-1}\right)^{-1}$$

Graph the imaginary part of this function to obtain a guess for the solution.

$$\omega := 7000, 7100 .. 8500$$

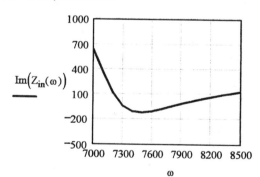

The two roots can be found using the root function.

$$\text{root}\left(\text{Im}\left(Z_{in}(\omega)\right), \omega, 7000, 7500\right) = 7.272 \times 10^3$$

$$\text{root}\left(\text{Im}\left(Z_{in}(\omega)\right), \omega, 7500, 8500\right) = 7.94 \times 10^3$$

15-37 Traditional, 16-37 Laplace Early. (A) TRANSFORMER TRANSFER FUNCTION

Figure P15-37 shows a transformer modeled as two coupled coils.

(a) Write the s-domain mesh-current equations for the circuit.
(b) Solve the mesh equations for the transfer function $V_2(s)/V_1(s)$. Show that the transfer function has a second-order bandpass characteristic.
(c) Find the center frequency, upper and lower cutoff frequencies, and bandwidth when $L_1 = 0.25$ H, $L_2 = 1$ H, $M = 0.499$ H, $R_1 = 50$ Ω, and $R_L = 200$ Ω . Does the transformer have a wideband or narrowband bandpass frequency response?
(d) Find the step response of the transformer using the element values in (c).

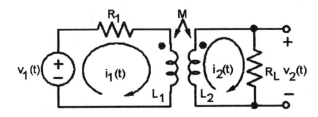

Figure P15-37

Solution: (a) The mesh equations produce the following matrix equation:

$$\begin{bmatrix} L_1 s + R_1 & -Ms \\ -Ms & L_2 s + R_L \end{bmatrix} \begin{bmatrix} I_1(s) \\ I_2(s) \end{bmatrix} = \begin{bmatrix} V_1(s) \\ 0 \end{bmatrix}$$

Which has the determinants

$$\Delta(s) = (L_1 L_2 - M^2)s^2 + (R_1 L_2 + R_L L_1)s + R_1 R_L \quad \text{and} \quad \Delta_2(s) = M s V_1(s)$$

We also have

$$V_2(s) = I_2(s)R_L = \frac{\Delta_2(s)}{\Delta(s)}R_L = \frac{M s R_L}{(L_1 L_2 - M^2)s^2 + (R_1 L_2 + R_L L_1)s + R_1 R_L}V_1(s)$$

So that

$$T(s) = \frac{M R_L s}{(L_1 L_2 - M^2)s^2 + (R_1 L_2 + R_L L_1)s + R_1 R_L}$$

(b) $T(s)$ can be rearranged to get

$$T(s) = \frac{\dfrac{M R_L}{L_1 L_2 - M^2}s}{s^2 + \dfrac{R_1 L_2 + R_L L_1}{L_1 L_2 - M^2}s + \dfrac{R_1 R_L}{L_1 L_2 - M^2}} = \frac{Ks}{s^2 + 2\zeta\omega_0 s + \omega_0}$$

(c) We use MATLAB for this part of the problem. The m-file and results follow:

```
% Define the components.
L1=0.25;
L2=1;
```

142

```
M=.499;
R1=50;
RL=200;
% Calculate the center frequency, bandwidth, damping coefficient, and
% upper and lower cutoff frequencies.
w0=sqrt(R1*RL)/(L1*L2-M^2)
B=(R1*L2+RL*L1)/(L1*L2-M^2)
zeta=B/2/w0
wC1=w0*(-zeta+sqrt(1+zeta^2))
wC2=wC1+B
%

w0 =   3.163859985841664e+003
B =    1.001001001001001e+005
zeta = 15.81929992920832
wC1 = 99.90029910309033
wC2 =   1.002000003992032e+005
```

Since ζ is much greater than 0.5, this is a broadband response.

(d) To find the step response we recall that
$$G(s) = \frac{T(s)}{s} = \frac{K}{s^2 + 2\zeta\omega_0 s + \omega_0}$$

In Mathcad we employ the Laplace inverse function to obtain the result

$$g(t) := \left(0.998e^{-t} - 0.998e^{-100099 \cdot t}\right) V$$

16-23 Traditional, 17-23 Laplace Early. The source in Figure P16-14 delivers 37 kW when the apparent power delivered to the load is $35 + j20$ kVA. The wire impedance is $2.1 + j12$ Ω. Find the load and source voltages.

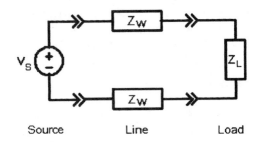

Figure P16-14

The MATLAB m-file and solution follow:

```
% Define the components.
PS=37000;
SL=35000+j*20000;
ZW=2.1+j*12;
% The Symbols are
syms VL IL VS
% The constraints are
f1=SL-VL*conj(IL);
f2=PS-real(VS*conj(IL));
f3=VL+2*ZW*IL-VS;
% Solution
[VL IL VS]=solve(f1,f2,f3)
```

results in

Warning: Explicit solution could not be found.

The solver in Matlab does not handle the complex values very well. I then created an m-file with 5 equations, separating the real and imaginary parts of the two complex equations.

```
% Define the components.
PS=37000;
SLre=35000;
SLim=20000;
ZWre=2.1;
ZWim=12;
% The Symbols are
syms VL ILre ILim VSre VSim
% The constraints are
f1re=SLre-VL*ILre;
f1im=SLim+VL*ILim;
f2=PS-VSre*ILre-VSim*ILim;
f3re=VL+2*(ZWre*ILre-ZWim*ILim)-VSre;
f3im=2*(ZWre*ILim+ZWim*ILre)-VSim;
% Solution
[VL ILre ILim VSre
VSim]=solve(f1re,f1im,f2,f3re,f3im,'VL,ILre,ILim,VSre,VSim');
VL=double(VL)
VSre=double(VSre);
```

144

```
VSim=double(VSim);
VS=VSre+j*VSim
% Check the solution.
PS=real(VS.*(ILre-j*ILim));
PS=double(PS)
SL=VL.*(ILre-j*ILim);
SL=double(SL)
%
```

results in

```
VL =
 -10.82663923921534
  10.82663923921534
VS =
  1.0e+003 *
  0.40924696324234 + 2.18671046034052i
 -0.40924696324234 - 2.18671046034052i
PS =
  1.0e+006 *
   4.04725384615385
   4.04725384615385
SL =
  1.0e+004 *
 -0.02051282051282 + 2.00000000000000i
 -0.02051282051282 + 2.00000000000000i
```

which do not check out. I then went to Mathcad to solve the original equations.

$$P_S := 37000 \qquad S_L := 35000 + j \cdot 20000 \qquad Z_W := 2.1 + j \cdot 12$$

$$\text{initial guesses} \qquad V_S := 2400 + j \cdot 10 \qquad V_L := 2400 \qquad I_L := \frac{V_S - V_L}{Z_W}$$

Given

$$V_L \cdot \overline{I_L} = S_L \qquad Re\left(V_S \cdot \overline{I_L}\right) = P_S \qquad V_S = V_L + 2 \cdot Z_W \cdot I_L$$

$$\begin{pmatrix} V_L \\ I_L \\ V_S \end{pmatrix} := \text{find}\left(V_L, I_L, V_S\right) \qquad \begin{pmatrix} V_L \\ I_L \\ V_S \end{pmatrix} = \begin{pmatrix} 1.847 \times 10^3 - 30.516i \\ 18.765 - 11.138i \\ 2.193 \times 10^3 + 373.068i \end{pmatrix}$$

Check the solution

$$V_L \cdot \overline{I_L} = 3.5 \times 10^4 + 2i \times 10^4 \qquad Re\left(V_S \cdot \overline{I_L}\right) = 3.7 \times 10^4$$

Obviously, this checks out.

16-29 Traditional, 17-29 Laplace Early. The load in Figure P16-29 operates at 60 Hz. With $\mathbf{V}_L = 12\angle0^{\circ}$ kV (rms) the load draws 1.2 MVA at 0.8 power factor lagging. The first source voltage is $\mathbf{V}_{S1} = 12.7 + j0.96$ kV (rms). Find the complex power supplied by each of the sources.

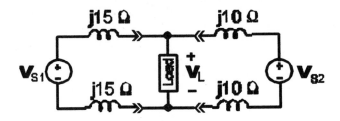

Figure P16-29

Solution: The MATLAB m-file and results follow:

```
% Define the givens.
VL=12000;
pf=0.8;
VS1=12700+j*960;
VA=1.2*10^6;
Z1=j*15;
Z2=j*10;
% The apparent power is then.
SL=VA*(pf+j*sqrt(1-pf^2));
% The load current is
ILconj=SL/VL;
IL=conj(ILconj);
% The current from source 1 is
I1=(VS1-VL)/(2*Z1);
% so that
S1=VS1*conj(I1)
% The current from source 2 is
I2=IL-I1;
% The voltage applied from source 2 is
VS2=VL+2*I2*Z2;
% so that
S2=VS2*conj(I2)
%

S1 =   3.840000000000000e+005 +3.270533333333333e+005i
S2 =   5.760000000000000e+005 +5.129688888888888e+005i
```

16-48 Traditional, 17-48 Laplace First. The three-phase load connected to Bus No. 2 in Figure P16-48 draws an average power of $P_2 = 100$ MW at a power factor of 0.8 lagging. At Bus No. 2 the Phase A voltage phasor is $90\angle0^O$ kV (rms). At the Bus No. 3 the Phase A voltage phasor is $90\angle12^O$ kV (rms). Find the complex power produced by each source.

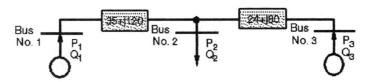

Figure P16-48

Solution: The MATLAB m-file and results follow:

```
% Inputs to the problem are
P2=100*10^6;
pf=0.8;
VP2=90000+j*0;
VP3=90000*exp(j*12*pi/180);
Z1=35+j*120;
Z2=24+j*80;
% Calculate S2
S2=P2+j*P2*sqrt(1/pf^2-1);
% Calculate the current from Bus 3
I3=(VP3-VP2)/Z2;
% The current into Bus 2 is
I2conj=S2/3/VP2;
I2=conj(I2conj);
% The current from Bus 1 is
I1=I2-I3;
% The voltage applied by source 1 is
VP1=VP2+I1*Z1;
% Finally the power supplied by each source is
S1=3*VP1*conj(I1)
S3=3*VP3*conj(I3)
%

S1 =   6.062877654549802e+007 +1.558675173023133e+008i
S3 =   5.976557428351042e+007 -1.129200600794657e+007i
```

16-51 Traditional, 17-51 Laplace Early. (A) UNBALANCED THREE-PHASE LOAD

A balanced three-phase source with $V_P = 208$ V (rms) drives a Y-connected three-phase load with phase impedances $Z_A = 100\ \Omega$, $Z_B = 100\ \Omega$, and $Z_C = 50 + j100\ \Omega$.

   (a) Calculate the line currents, line voltages, and total complex power delivered to the load.

   (b) Repeat part (a) when a zero-resistance neutral wire is connected between the source and load.

   (c)

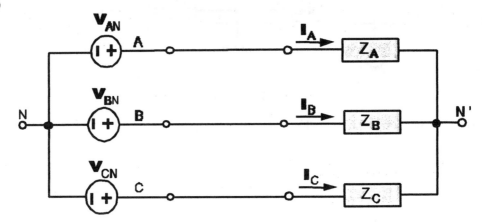

Figure P16-51

Solution: The MATLAB m-file and results follow:

```
% The inputs are
VP=208;
ZA=100;
ZB=100;
ZC=50+j*100;
% The relevant voltages are
VAN=VP
VBN=VP*exp(-j*120*pi/180)
VCN=VP*exp(-j*240*pi/180)
VAB=VAN-VBN
VBC=VBN-VCN
VCA=VCN-VAN
% From the node equation at N' with point N grounded we have
VN=(VAN/ZA+VBN/ZB+VCN/ZC)/(1/ZA+1/ZB+1/ZC)
%  From which we get
IA=(VAN-VN)/ZA
IB=(VBN-VN)/ZB
IC=(VCN-VN)/ZC
SL=VAN*conj(IA)+VBN*conj(IB)+VCN*conj(IC)
SLre=real(SL)
% With point N' grounded, VN=0 and hence:
IA0=VAN/ZA
IB0=VBN/ZB
IC0=VCN/ZC
SL0=VAN*conj(IA0)+VBN*conj(IB0)+VCN*conj(IC0)
SL0re=real(SL0)
%

VAN =  208
VBN = -1.040000000000000e+002 -1.801332839871633e+002i
VCN = -1.040000000000001e+002 +1.801332839871632e+002i
```

```
VAB =   3.119999999999999e+002 +1.801332839871633e+002i
VBC =   1.278976924368180e-013 -3.602665679743265e+002i
VCA = -3.120000000000001e+002 +1.801332839871632e+002i
VN =   80.54998149518620 +16.48333950160461i

IA =    1.27450018504814 -  0.16483339501605i
IB =  -1.84549981495186 -  1.96616623488768i
IC =    0.57099962990372 +  2.13099962990372i

SL =    1.135680000000000e+003 +4.867200000000001e+002i
Slre = 1.135680000000000e+003

IA0 =   2.08000000000000
IB0 = -1.04000000000000 -  1.80133283987163i
IC0 =   1.02506627189731 +  1.55253313594865i

SL0 =    1.038336000000000e+003 +3.461120000000001e+002i
SL0re = 1.038336000000000e+003
```